Chinese economic reform

As a result of intensive economic reform and open door policies in the past fifteen years, China has become a major player in the global economy. This economic prosperity has brought great changes to the country, and in particular to the military, who, with new resources available, have redefined China's internal and external security policies.

Chinese Economic Reform reaches beyond the economic success of recent years to focus specifically on the pivotal role of the People's Liberation Army (PLA), and points towards its vigorous economic participation as a means of consolidating power. The contributors address a wide range of topics, including the PLA's business activities, military industry and conversion, and arms sales. These studies highlight the way in which the military now acts as a business in this new environment, with its own form of insider dealing and large-scale corruption. Inevitably, these changes have affected China's relationship with the rest of the world; this volume carefully evaluates the effects of rapid economic growth on China's policies towards defence up to the end of the twentieth century.

Gerald Segal is a Senior Fellow at the International Institute for Strategic Studies, London. **Richard H. Yang** is Emeritus Professor of Washington University, USA, and the Chairman of the Chinese Council of Advanced Policy Studies, Taipei, Taiwan.

Chinese economic reform

The impact on security

Edited by
Gerald Segal
and
Richard H. Yang

Published for the
International Institute for Strategic Studies
and the
Chinese Council of Advanced Policy Studies

London and New York

First published 1996
by Routledge
11 New Fetter Lane, London EC4P 4EE

Simultaneously published in the USA and Canada
by Routledge
29 West 35th Street, New York, NY 10001

Routledge is an International Thomson Publishing company

Typeset in Times by
Ponting–Green Publishing Services, Chesham, Bucks

Printed and bound in Great Britain by
TJ Press (Padstow) Ltd, Padstow, Cornwall

British Library Cataloguing in Publication Data
A catalogue record for this book is available from the
British Library.

Library of Congress Cataloguing in Publication Data
Segal, Gerald, 1953–
Chinese economic reform: the impact on security / edited by Gerald Segal and Richard H. Yang.
p. cm.
Includes bibliographical references and index.
1. Military readiness–Economic aspects–China.
2. Defense industries–China. 3. China–Economic policy–1976– 4. National Security–China. 5. China. Chung-kuo jen min chieh fang chün. 6. China–Foreign relations–1976–
I. Yang, Richard H. II. Title.
HC430.D4S44 1996
338.951–dc20 95–24664
CIP

ISBN 0–415–13007–7

Contents

Notes on contributors

Arthur S. Ding is an Associate Research Fellow at the Institute of International Relations of the National Chengchi University in Taipei, Taiwan. His research focus is on the Chinese People's Liberation Army, and he has written extensively on the PLA's perception of the world, military strategy, defence conversion, and business activities and defence budget.

François Godement is Professor, Chair of Chinese Civilization, at the National Institute of Oriental Languages and Civilizations (INALCO), Paris, Director of International Trade Department of INALCO (since 1986), and Senior Research Associate, in charge of Asia–Pacific affairs, French Institute of International Relations, Paris (since 1986). His most recent publication is *La Renaissance de L'Asie* (Paris, Odilo Jacob, 1993), to be published in English by Routledge.

Paul H. B. Godwin is Professor of International Affairs at the National War College, Washington, DC. His research focuses on Chinese security policy and defence modernization. His most recent publication is 'China and Arms Control: Transition in East Asia,' *Arms Control Today*, vol. 24, no. 9, November 1994. Other essays have appeared in *Studies in Comparative Communism, Contemporary China, The China Quarterly* and *Armed Forces and Society.*

David S. G. Goodman is Director of the Institute for International Studies, University of Technology, Sydney. His recent publications include *Deng Xiaoping and the Chinese Revolution* (Routledge, 1994); with Beverley Hooper, *China's Quiet Revolution* (Longman, 1994); and with Richard Robison, *The New Rich in Asia: Mobile-phones, McDonalds and Middle Class Revolution* (Routledge, 1995). With Gerald Segal he has recently published *China Deconstructs* (Routledge, 1994) and *China without Deng* (ETT, 1995). He has recently completed a study of political change in the PRC, *Transformations in the Chinese State: The Reform Agenda and Regime Change.*

Ellis Joffe is Professor of Chinese Studies at the Hebrew University. He was born in Shanghai, China, and educated in Jerusalem and at Harvard University. He has been a research fellow in Chinese Studies at Harvard, the University of Michigan, the Australian National University, and St Antony's College, Oxford University. His latest book is *The Chinese Army After Mao.*

Michael Leifer is Professor of International Relations at the London School of Economics and Political Science. He has held visiting appointments at the Australian National University, the University of the Philippines, Cornell University and the National University of Singapore. He has conducted research on East Asia and especially Southeast Asia for many years and his most recent book is *A Dictionary of the Modern Politics of Southeast Asia* (Routledge, 1995).

Masashi Nishihara is Director of the First Research Department of the National Institute for Defence Studies, Tokyo. He is also Professor of International Relations at the National Defence Academy. He is currently a Research Associate of the Research Institute for Peace and Security, Tokyo. Professor Nishihara also serves on the Council of the International Institute for Strategic Studies, London.

Gerald Segal is a Senior Fellow at the International Institute for Strategic Studies and Director of the Economic and Social Research Council's *Pacific Asia Programme.* He has held senior appointments at the University of Bristol and the Royal Institute of International Affairs and is a regular contributor to the media on Asian affairs. Dr Segal has written or co-authored more than a dozen books or monographs and has edited or co-edited a similar number. His most recent publications include *The World Affairs Companion* (Simon & Schuster, 1993), *The Fate of Hong Kong* (Simon & Schuster, 1993) and *China Changes Shape* (*Brassey's* for the IISS, 1994).

Michael D. Swaine is a Senior Political Scientist in the International Policy Department of RAND in Santa Monica, California, and Co-director of the RAND Center for Asia–Pacific Policy (CAPP). Dr Swaine has worked at RAND since 1989. He was formerly a research associate at Harvard University, a Postdoctoral Fellow at the Center for Chinese Studies, University of California, Berkeley, and a project manager and consultant in the business sector.

Michael Yahuda is a Reader in International Relations at the London School of Economics. He has held visiting professorships at the University of Adelaide and the University of Michigan (Ann Arbor). He is the author of several books and many scholarly articles on China's foreign relations and the international politics of Asia and the Pacific.

Richard H. Yang is Professor Emeritus of Chinese Studies at Washington University, USA. He currently teaches at National Sun Yat-sen University and is Chairman of the Chinese Council of Advanced Policy Studies (CAPS). He is the author of many books and articles and editor of *Chinese Regionalism: The Security Dimension* (Westview, 1994).

Acknowledgements

This book contains the proceedings of an international conference held in Hong Kong on 8–10 July 1994 on China's military affairs. The conference, the second of its kind, was jointly organized by the Chinese Council of Advanced Policy Studies (CAPS) in Taipei and the International Institute for Strategic Studies (IISS) in London. Participants came from the Hebrew University of Jerusalem; University of Technology, Sydney; National War College, Washington, DC; London School of Economics (LSE); School of Oriental and African Studies (SOAS), University of London; American Enterprise Institute (AEI); RAND Corporation; Miami University; Institute of International Relations, National Chengchi University; International Institute for Strategic Studies; and Chinese Council of Advanced Policy Studies.

Essential administration and financial support for the conference was given by the Chinese Council of Advanced Policy Studies. Special thanks are given to Gerald Segal, Andrew N. Yang, Peter K. H. Yu, Sheelagh Urbanoviez and Yi-Su Yang for their important contribution in planning and assistance in conference arrangement and their immense effort for making this book possible.

Richard H. Yang

1 Introduction

Gerald Segal and Richard H. Yang

The People's Republic of China (PRC) often complains that it is misperceived by the outside world. One important reason for the difficulty in perceiving China is the fact that China, willingly or unwillingly, projects a number of different images. As a self-proclaimed Communist Party state, it is not surprising that many in the non-Communist outside world will be suspicious of Chinese motives. As a state that avowedly places greater virtue on stability than pluralist politics, it should also not be surprising that the outside world takes a different view of Chinese human rights policy. There is nothing intrinsically wrong with any of these different perceptions, they may simply mirror different realities and interests.

What is more difficult to reconcile is conflicting realities and interests. These are most apparent when considering at least two other priorities of Chinese policy. China is avowedly not a status quo power. It openly proclaims its desire to regain control of territory and people lost in earlier days when China was weak and divided. Thus, China claims the right to regain islands in the South China Sea, the people of Hong Kong, or the more numerous people on Taiwan. Not surprisingly, other states contest territorial claims and other people contest China's ability to determine their fate. These differing interests may or may not be reconcilable, but they are important and currently unreconciled.

These non-status-quo objectives also do not sit easily with another Chinese objective, that is the determination to make China rich by reforming its economy and trading with the outside world. As China emerges as a major player in the international economy, important questions are raised about how the existing international economy should adapt or even change its rules. Is China a non-status-quo power in economic terms? If it says it is willing to play by existing rules, then how much will its other objectives be constrained by the needs to play by the rules of the international economy?

These are difficult questions for any emerging power, let alone one the size of the PRC. They are made even more complicated by the fact that these seemingly international issues are in fact tied up with fundamental issues in Chinese domestic affairs. As China reforms its economy, it finds it needs to change aspects of its political system and even the way in which its armed forces operate. By placing economic reform at the top of its agenda, the leaders of the PRC have ensured that they will have to resolve these issues, at least temporarily.

Understanding the impact of economic reform has become a growth industry for the China studies community. Remarkably, this industry has not produced a book-length study of the impact of Chinese economic reform on defence policy. Part of the reason might well have had to do with the natural conflicts of policies posed by economic reform. Another part of the problem might have to do with the difficulty of gathering experts with the ability to cross the disciplinary boundaries of defence studies and economics. But despite these difficulties, the subject was deemed so important that the International Institute for Strategic Studies (IISS) in London and the Chinese Council of Advanced Policy Studies (CAPS) in Taipei decided to pool their resources in order to tackle the subjects.

The IISS and CAPS collaborated a year earlier in assessing the equally vexed question of Chinese regionalism. In both cases we chose to hold a multi-disciplinary meeting in Hong Kong in order to bring together the best possible gathering of relevant expertise from around the world. Experts were drawn from more than a dozen countries and the critical audience included specialists in a wide range of disciplines. The meeting on Chinese economic reform and defence policy took place in July 1994 and, like its predecessor a year earlier, provoked much comment in the regional and international media. At the time of the meeting, China's leaders were particularly anxious about their ability to exercise control over the economy and, of course, since Sino-British relations regarding Hong Kong were tense, the Chinese authorities were particularly sensitive to a meeting about defence and economic policy in Hong Kong. For the conference participants, the result was a lively and well-focused debate.

The revised papers from the meeting are published in the hope that they will contribute to the wider debate. By the time they are published, the specific headlines will have changed, but the essentials of the problems raised in the papers are likely to remain current. We offer these papers in the expectation that they will help people find ways to think about the implications of Chinese economic reform for defence policy. What follows is a brief outline of the essential arguments and evidence, without attributing any specific conclusion to specific chapters or commentators.

INTERNAL IMPACT

Previous conferences sponsored by CAPS, and indeed much of the Chinese civil–military relations field, have been dominated for years by discussions of the precise balance between Red and Expert. Although opinions differed, there was regular assent to the notion that professional military men resented interference from the civil sector. There was also general assent that the direction of civil–military relations in China in recent years has been towards greater professionalism. What the specialists rarely had to confront was the fact that professionalism among soldiers could be undermined by something much more powerful than politics–money. Where decades of political campaigns had failed to undermine professionalism, a few years of economic reform had eaten into the core of the People's Liberation Army (PLA) values.

The extent to which money has corrupted professionalism is still uncertain. Anecdotal evidence is plentiful that large parts of PLA activity are now concerned with moneymaking enterprises. Some suggest that up to half the PLA's order of battle is now concerned with economic activities. To be sure, the PLA was always engaged in non-military economic activities. The difference is now the scale of such operations, and the fact that so much of it is for profit, and often the profit of individuals and/or their units. Optimists suggest that such activity is merely a prelude to privatization and will soon be good for PLA professionalism. Pessimists argue that, at a minimum, such a dual structure in the PLA saps morale and professionalism. If the optimists are wrong, in the long term no military establishment can survive the cancerous appeal to moneymaking without serious damage to professional skills. If we are lucky, the PLA is on the road to privatizing half its order of battle, thereby reducing the threat to its neighbours. In so doing, the PLA may also emerge as an institution with a major stake in continuing economic reform.

One of the major features of economic reform has been the extent to which economic power has been decentralized, not just to the provinces, but to a wide range of entrepreneurs and organizations. This, of course, was the topic of the IISS–CAPS conference in June 1993, but the subject was revisited in more concentrated form in 1994 because of the importance of regionalism in economic reform. As was concluded in 1993, the threat of 'warlordism' is not critical, although in the past year there has been a greater development of more regionally specific interests within the PLA as within the society as a whole. The failure of the central government to get a grip on the economy, most notably after the failed reform of November 1993, was also accompanied by increased signs of decentralization in the armed forces. The persistence of stories about freelance activities by military units, especially in border and coastal regions, makes this point most vividly. In short, so long as the society, economy and political system show important signs of decentralization, so the armed forces can be expected to exhibit similar tendencies. The risks to Chinese unity and hence to defence policy are potentially severe.

The risks from decentralization were not thought too severe when the economic reforms were first begun. At that time, the PLA feared the fact that its modernization was ranked fourth of the Four Modernizations. The PLA had been promised that once the economic pie grew, they too would benefit. Indeed, with average growth rates of 9 per cent in the age of reform, even a formally declining percentage of defence spending meant in reality an increase in defence spending in many years. After the events of 1989, even the percentage of state spending on defence rose, and the PLA began to experience real increases in defence spending. It was remarkable how long it took the China-watching community to absorb the reality of de facto increases in defence spending. Acknowledgement of these trends, and analysis of their nature, only began in the past few years.

Calculation of Chinese defence spending remains contentious, if only because of the poor base of information. The chapters in this volume reach contradictory conclusions, as does work by all the major institutes studying the subject. But some trends are evident and agreed. Most notable is the conclusion that with the upward revision of the size of the Chinese economy based on purchasing power parities, so the defence budget has to be revised upward. There is also agreement on the fact that the real spending on defence is far higher than the stated budget. Finally, there is also agreement that if the Chinese government would cooperate more fully with its neighbours in providing more complete data on its defence expenditure, then there might be less worry about Chinese intentions. In the absence of harder data, defence planners in other countries have an institutional responsibility to be cautious and worry about signs of increased and increasing Chinese defence spending.

Another way to understand the nature of the risk posed by increased defence spending is to assess what is taking place in aspects of the Chinese defence industry. This task is also fraught with severe problems of poor data. China has made much of its conversion of industry from military to civil products, but few analysts take their data seriously. Much concern exists about the degree of profitability of defence industry and where any profits might go. So much of the defence industrial sector is said to be caught up in the problems of corruption and false accounting that are evident elsewhere in the state-owned industry, that it is hard to reach firm judgements.

The Chinese remain aware of the challenges of new technology and the need to update their defence industry. Yet they must also be aware, although rarely discuss it, that every other great power is cutting its defence industry. China happens to be in a region where defence spending is sharply increasing in real terms, although not as a percentage of total spending. The result is that China is no longer in an arms race, however tentative, with the great powers, but rather with its neighbours. The impact of increased defence spending and modernization of defence industry is felt most keenly in East Asia rather than among the great powers. It is true that Europeans and Americans tend to be the most vocal in their criticism of Chinese defence spending, but they stand to suffer less if the trends continue. The implication is that at some point China will have to answer to its neighbours about its defence economic policy, or else risk a further deterioration in East Asian security.

EXTERNAL IMPACT

If the dominant sense of the internal impact of economic reform on defence policy is one of weakening central control and yet increased national power, not surprisingly much the same is true when considering the external impact. And yet there are important trends in the world outside China that make for different nuances in the impact of economic reform. Of course the starting point must be that economic reform has made China a major trading power. Its trade surpluses with the United States, the EU and now Japan, suggest that it will run structural surpluses with developed states as Chinese exports sell in these markets. Much of China's economic reform depends on its external orientation and hence its access to foreign markets. The result is critical dependence by China on these markets and therefore critical leverage on the part of those who control the gateways to those markets.

A second feature is Chinese dependence on imports, especially of food and fuel. Economic growth has helped drive parts of China, and especially coastal regions, to develop specialized ties with foreigners. Southern China sees the economic sense of dependence on Australian food or fuel rather than having to buy either more expensively from

Northeast China. This is the essence of complex interdependence in the modern economy, and it has a major impact on China's geo-economic position. Interdependence suggests dependence and dependence suggests limits on defence and security policy.

Such a series of logical links is what leads many of China's neighbours to take a more benign view of Chinese security policy. Their assessment is that China is dependent on access to the international economy and therefore there is less reason to worry about how it disposes of its troops. The calculation is credible, but perhaps not cautious. China's dependence is clear, but it is less clear that China draws the same conclusions. Dependence can also lead states, as in the case of Japan earlier in the century, to seek ways to reduce dependence by controlling sea lanes and resources, as well as by cowing neighbours. In short, there is nothing inevitable about accepting the constraints of interdependence. This is especially so when China feels that it has a status quo to re-order and has a heritage of dominance in the region.

The evidence about Chinese behaviour in the recent years of growing interdependence is not reassuring. China's policy towards Hong Kong has shown it is unwilling to sacrifice nationalist goals for economic gain. Quite the reverse in fact. Similar tendencies are evident in China's relationship with Taiwan. So far there is not a single case where China has muted or abandoned a nationalist objective for the sake of economic gain. In short, there is no evidence, yet, of interdependence constraining Chinese behaviour.

On the other hand, there is plenty of evidence of China's neighbours constraining their own behaviour for fear of putting at risk their economic links with China. The message that such self-restraint sends to Beijing is that interdependence makes others dependent on China and not the other way around. For proud Chinese with an historical chip on their shoulder, this is seen as a natural state of affairs.

Of course, not all countries have reacted to Chinese power in the same way. The USA and the EU have also struggled with different strategies for dealing with China, and still have not found a happy resting place. For a time after the events of 1989 it was felt that pressure on human rights could succeed in restraining China, especially when foreigners threatened to restrict Chinese trade unless human rights were observed. In a short time, the Western powers abandoned such a strategy, in part because it was complex to implement and hard to prove its efficacy. Most importantly, Western business interests pressed for a relaxation of the policy.

But it was other parts of the business community that supported the tough stand that created the most successful example of Western pressure on China. It did not escape Western businessmen for long that China was running a huge trade surplus with their countries and therefore there were likely to be some unfair trade practices on China's part. Nowhere has the West been more successful than in its ability to force China to abide by international trade accords. It has done so by threatening to withdraw access to Western markets. The debate about Chinese entry into the General Agreement on Tariffs and Trade (GATT) concerned how much China would have to reform its behaviour before it was allowed into an existing institution. East Asian countries have argued for greater lenience in its relations with China, but the Europeans and Americans have less reason to appease China. This conflict of interests with China, and among states who deal with China, is set to continue.

What is also set to continue is a concern with geoeconomics when trying to understand Chinese defence policy. Chinese policy in the South China Sea is in part driven by concern to secure energy and mineral resources. Relations with Taiwan are in part affected by the growing economic connection across the straits. Economics is also part of the explanation for China's arms sales and especially its concern with Middle Eastern markets. As China becomes a major oil importer, it will have to find something to sell to the Gulf states in exchange. Even the very growth of the Chinese economy itself causes concern for the likes of Japan. The one area that Japan used to dominate was in regional economic power, and now that looks set to fade. When we talk of Chinese power in the future, it will be as much in economic terms as in military terms.

ADJUSTING THE AGENDA

It is a truism of strategic studies that after the Cold War the job description changed. Much the same can be said for China specialists, and especially for those who have specialized in the study of the PLA. This is not to say that the old agenda is entirely without value, but rather that new skills need to be learned if analysts are to understand their subject.

Students of Chinese politics have already made good strides in learning about civil society and how it develops in post-Communist states. More recently there has been some recognition that the skills of Beijing watchers are no longer sufficient when they should be watching Shanghai or Guangzhou. Few specialists on the PLA have looked at how these issues affect the armed forces, but there is some progress in filling analytical gaps.

But by far the largest challenge is to understand the economics of defence policy. It is remarkable how little economic literacy there is when the subject is as basic as the role of inflation or the meaning of purchasing power parities. Students of Chinese defence policy need to spend a little less time studying the latest bit of kit, if this means they will have a bit more time to understand trade balances and financial flows. Of course, part of the difficulty in acquiring such knowledge is that it requires the ability to cross well entrenched lines of discipline. But unless the effort is made, one fears that the analytical record of Sinologists will continue to lag well behind that of meteorologists.

Part I

The impact on internal defence policy

2 The PLA and the economy

The effects of involvement

Ellis Joffe

After the October 1992 Party Congress, it seemed that the future development of the People's Liberation Army (PLA) would be influenced by military professionalism much more than in any past period. By that time, the political assault on the PLA, which had been launched following its uncertain behaviour during the Tiananmen crisis, had run its course, and PLA activities had been refocused on improving military capabilities. Lending impetus to this effort was the shock given to the Chinese leadership by the Gulf War, which highlighted the awesome complexity of high-tech warfare and spotlighted the chasm that separated the PLA from advanced armies.

The Fourteenth Congress put the seal on the new ascendance of professionalism in the PLA: it dismissed Yang Baibing, the director of the General Political Department and the nemesis of the professional military, and appointed two veteran advocates of military modernization, Liu Huaqing and Zhang Zhen, to head the armed forces. The PLA seemed set on a course that would be shaped above all by military concerns.

Or was it? Just when political intrusion into military affairs had been drastically reduced, a new threat to the professional integrity of the PLA seems to have appeared: military intrusion into economic affairs. This intrusion has been the direct result of China's economic reforms.

Of course, the involvement of the military in economic affairs is nothing new. It began in the early days of the Red Army, expanded during the revolutionary period, and continued, with widely varying intensity, after the establishment of the Communist regime and until the end of the Maoist period. But following the start of Deng's reforms, the nature of this involvement was radically transformed. As the Deng leadership cast off Maoist restraints on economic development, the PLA began to move into commercial activities that would have been strictly off-limits on ideological grounds during the Maoist period. These activities expanded rapidly during the 1980s and underwent a dramatic intensification in the wake of Deng's southern tour of early 1992, during which he called for an acceleration of capitalist type economic reforms.

No one will argue that these unconventional economic pursuits are potentially damaging to the PLA as a military organization. They pose a danger to its combat capability, professional ethic, internal cohesion, and subordination to the central leadership. The question is whether these pursuits have already affected the PLA and what is the outlook if it continues to be preoccupied with making money. In short, is the PLA changing its characteristics and becoming an army whose professional skills are being eroded by its economic activities? If it is, as some observers argue, the implications are enormous for its military posture and political role.

The purpose of this chapter is to examine how far the PLA has moved in this direction and what the prospects are for the future. It will try to answer several questions: what are the dimensions of the PLA's economic involvement? What have been its effects on military capability and cohesion? On civil–military relations? What has been the response of the leadership? What is the balance sheet?

WHAT IS ECONOMIC INVOLVEMENT?

An assessment of the effects produced by the economic involvement of the military must begin by inquiring into the types of activities carried out by the PLA and their scope. This inquiry is essential because generally there is no immediate and visible link between these activities and their effects. The link has to be worked out by deduction and speculation, for which a familiarity with the variety and volume of the PLA's economic involvement is imperative.

The problem is that it is extremely difficult to get information on this involvement. First, political and military leaders are not interested in publicizing many of the PLA's economic activities because these conflict with the army's traditions and desired image. Second, military units evidently try to conceal many of their economic ventures because they are illegal, or because unit commanders want to keep the profits for themselves. Third, it is obvious that no central organ, political or military, has precise data on the PLA's far-flung involvement in economic affairs, and the Chinese could not provide this data even if they wanted to. Thus, the base line for evaluating the effects of economic reforms on the PLA can only be drawn in a sketchy and tentative fashion.

The first step is to define what is meant by the term 'economic involvement' of the PLA. This is necessary because there is a tendency to blur the distinction between two categories of activities which have entirely different effects on the armed forces. This tendency may result in a distorted picture of actual PLA involvement in the economy and its consequences.

The first category consists of economic activities that have a military dimension, but which are not carried out by the PLA. The main area of activity in this category is the production of civilian goods by military industries. While these activities are frequently portrayed as indicating the intrusion of the PLA into the civilian economy, this, in fact, is not so. For one thing, the military industries are not under the control of the PLA. More important, their activities are removed from the daily tasks of the PLA and do not involve the participation of its members. If these activities affect the PLA, this is likely to occur in one way only – by increasing its budget. And it is not at all certain that this has, in fact, occurred. On the other hand, the second category encompasses a range of economic activities which employ large numbers of PLA members on active service and which are likely to have direct and significant effects on its character as a military organization. A look at these different categories will help clarify the nature of the PLA's economic involvement.

THE CONVERSION OF MILITARY INDUSTRIES

First, the military industries. From the time they were established in the early 1950s, these industries devoted a portion of their production to the civilian market, but until 1979 its value was negligible – not more than about 8 per cent of total production.[1] In that year a major change took place, when the Chinese leadership decided to convert these industries to civilian production. Conversion started out slowly but picked up rapidly from the mid-1980s, so that by the early 1990s between 66 and 76 per cent of total production by military industries were reportedly designed for the civilian market.[2]

The reasons that underlay the decision to convert military industries also explain why the shift has been extensive. One reason was the launching of Deng's reforms, which created a pressing need for advanced technologies and specialists in various sectors of the economy – assets that were readily available in the military–industrial complex.

These assets could be diverted to the civilian sector because of a dramatic change that occurred at that time in the threat perception of Chinese leaders. After being obsessed with the fear of a Soviet attack for a decade, the Chinese began to adopt an increasingly relaxed view of the Soviet Union, due to the waning of Soviet global power that resulted from problems at home and the build-up of American military might. This view obviated the need for urgent weapons procurement and enabled the Chinese to place economic development ahead of military modernization and to channel the output of military industries to civilian demand.[3]

Such a shift was strongly encouraged by the stark fact that a considerable portion of the weapons and equipment produced by these industries was outdated. When Mao's doctrine of 'people's war' – which pitted quantity against quality – dominated the planning for a future conflict and the development of the armed forces, perhaps there was a certain rationale for producing large quantities of backward weapons. However, as the armed forces began to modernize and Maoist doctrine was dismantled, this rationale gave way to a new approach. Formulated by the mid-1980s, this approach postulated that China no longer faced a major war in the foreseeable future but had to prepare for limited local wars that required smaller quantities of more sophisticated weapons.

As demand for weapons decreased, what stood out was the excess capacity of the military industries, which reportedly consisted of about 2,000 enterprises, with hundreds of affiliated research institutes, which employed some three million workers. Among the largest were those that belonged to the so-called 'third line'. Built from 1964 for strategic purposes, these industries were scattered throughout China's inland provinces as a precaution against an attack. They totalled about 25 per cent of all military enterprises and absorbed about one-third of their work force. They were huge self-sufficient units, which had vast resources at their disposal, and which were far ahead of many civilian enterprises in the level of their technology and the skills of their personnel. However, in the new situation of dwindling need for military products, the enormous capacity of these industries became partly superfluous. At the same time, as the reforms progressed, the demand for civilian products, which these industries were capable of manufacturing, increased. It was only natural, therefore, to turn their production to the civilian sector. This symbiosis gave new life to China's military industries.[4]

A few details will illustrate the scope of their activities after conversion. According to official reports, which are uniformly glowing, since 1979 the military industry has produced more than 700 kinds of civilian goods, which include cars, motorcycles, freight trains, train axles, refrigerators, cameras, chemical products, and firearms for civilian use.[5] Also included are aircraft, tankers, and container ships, as well as various kinds of household appliances, textiles, foodstuffs, and medicines.[6] One-fifth of the cameras produced in China, 65 per cent of the motorcycles, and three-fourths of the passenger vans that move in the streets of Beijing – all these come from the military industries. Radiation technology from this sector has reportedly been used to cultivate new breeds of farms products and to increase grain and cotton output. The world's first slanting suspension bridge is said to have been built by the defence sector[7] – and the list goes on and on.

As a result of these activities, the output value of civilian goods rose by an average of 25 per cent from 1979 to 1989.[8] More than 100 types of products were exported in 1989 with a value of more than $300 million.[9] And some military enterprises have established joint ventures with foreign companies and their products, such as passenger aircraft, are exported.[10]

To what extent these industries have also been profitable is a moot point. On the one hand, their reported successes suggest that they have been raking in a lot of money. On the other hand, however, conversion has clearly not been an easy process and many industries have failed – so far, at least – to adjust to the new circumstances and have yet to turn a profit. For example, one huge conglomerate, the China North Industries Corporation (NORINCO) has had only limited success in conversion due to the difficulty of shifting its tank and artillery plants to civilian production. As a result, NORINCO has been losing money.[11] So, it may be assumed are other converted military industries which are still in need of state subsidies. The most that can be said on balance is that some industries are probably making money while others are not.

Whether some of the revenues from the profitable industries find their way to the PLA is not known. Given the subordination of the arms industries to the Commission on Science, Technology and Industry for National Defence and, ultimately, to the Central Military Commission – which is also responsible for the PLA – and given the personal ties between the heads of the military–industrial bureaucracy and the armed forces, it is reasonable to assume that part of their profits may have been channelled to the PLA. According to one Hong Kong report, in 1992, 65 per cent of the profits made by the military industries through civilian production was added to the military budget.[12] However, even if true, such an allocation was probably an exception because in that year there was a particular need occasioned by the purchase of Russian aircraft and surface-to-air missiles.

Even if the PLA has received funds from the converted defence industries, it has been a passive recipient. Not only are uniformed PLA officers and men not employed in this sector, but the heads of the three ministries in charge of the arms industry are civilians.[13] So apparently are management personnel in the military factories. For example, leadership positions in one enterprise visited by a foreign observer in 1989, the Xi'an aircraft factory, were held by technical professionals, very few of whom were former, not current, PLA officers.[14] In short, the PLA is institutionally separate from the defence industries. Therefore, the production of civilian goods by military industries is not tantamount to PLA involvement in the economy. And yet, the PLA is very much involved – but in another category.

DIRECT PLA INVOLVEMENT

The involvement in this category is direct, far-reaching, and many-sided. However, it is completely different from the well-known traditional participation of soldiers in non-military activities, which has been a marked feature of the Chinese Communist army from its inception until the present. During the revolutionary period this participation grew out of two related needs that were dictated by the circumstances in which the Red Army operated. One was the need to ease the burden of the army's upkeep on the population in order to reduce the friction inevitably caused by the widespread dependence of the troops on food grown by the farmers. The other was the need to gain the goodwill of the people in order to ensure the cooperation without which the Red Army could not survive.

These needs determined the types of activities in which the army engaged. They included the raising of crops and livestock for the consumption of the troops; running small industries which catered to the basic needs of the soldiers; helping peasants in routine activities, such as harvesting, and in times of trouble, such as floods. The extent to which the army actually carried out such activities varied widely in time and place, but there is no doubt that its commitment to helping the people was a major factor in gaining their support and, therefore, in bringing the Chinese Communists to power.

After the establishment of the People's Republic, Chinese leaders did not abandon the principle of PLA participation in non-military activities. On the contrary, to this day they highlight the importance of these activities to preserving good relations with the population and to reducing the costs of maintaining the armed forces. However, the application of the principle differed greatly in line with prevailing circumstances.

When Maoist doctrine dominated the daily life of the PLA, and this covered almost the entire period from 1958 to 1976, there were few military restraints on the employment of troops in external tasks.

However, the actual employment was probably never nearly as extensive as the regime's propaganda depicted it, although it was extensive enough to arouse the opposition of professional military leaders. When these leaders gained the upper hand in shaping military policy after the end of the Maoist period, the army's outside activities were sharply reduced – regardless of the pictures painted by the media – at least to a point which was acceptable to military commanders. But whatever their extent, one essential feature was common to all these activities: they were not carried out for the purpose of making money. And this is what distinguished them from the pursuits-for-profits which have become a novel and significant fixture of the Chinese military scene.

The PLA began to branch out into these pursuits as a result of two factors which coalesced in the early 1980s, and which account for their great expansion since then. One factor derived from the military policy of Deng and his colleagues. Its essence was that, although modernization of the PLA was designated as one of their main objectives, they ruled out the rapid renovation of weapons and equipment because this would have been impossibly expensive. Instead, they decided that the technological upgrading of the PLA would be a slow process that would be preceded by China's economic and technological advancement.

This policy has shaped the military budget for more than a decade. And, for the most part, it was a lean decade. The stated military budget did not change much between 1980 and 1988, rising from about 19.4 billion yuan to about 21.8 billion yuan RMB (Renminbi). However, military spending as a portion of total expenditures dropped drastically – from about 16 per cent in 1980 to about 8.2 per cent in 1988. The second half of the 1980s was the hardest in terms of military spending.

The military have fared much better in the 1990s. In 1990, the budget was raised by about 9 per cent, mainly as a result of the Gulf War, and by another 12 per cent in 1991. In 1992 the budget increased by about 14 per cent and in 1993 by another 15 per cent. In 1994 it was raised by about 23 per cent. However, despite what appear to be major increases in military appropriations in the past few years, according to military officials the actual purchasing power of the PLA has not increased. The biggest deficits were between the mid to the late 1980s.[15]

The shortage of funds created major difficulties for the PLA. Although professional military leaders were placated to some extent by the leadership's sanction, indeed encouragement, to implement sweeping military reforms in areas that did not require money, the shortage of funds has been a prime source of dissatisfaction. Not surprisingly, the Chinese military complained time and again that, as an article in the army paper put it, limited military spending can hardly satisfy the needs of army operation and development. 'This is the really serious situation our Army is facing.' This situation was most serious in weapons acquisition: 'the greatest obstacle to our Army's efforts to modernize its armament and equipment is the shortage of funds'.[16]

Another difficulty was that there was not enough money for the upkeep of the armed forces. For example, according to one estimate, the military budget for 1992 (for current expenses only) should have been RMB 50 billion, but in fact was only RMB 37 billion.[17] The situation in other years was similar. This was made clear, for example, by officers who were in the best position to know. In 1987, the deputy director of the PLA's General Logistics Department said that 'the defense budget this year has a slight increase over last year. However, after allowing for price rises, defense expenditures have not actually increased'. Things did not get better two years later. As described by the director of the Department: 'The state has made considerable allowance for military expenditure for 1989 despite financial difficulties. But the army will remain short of money in 1989 due to price increases, new expenditures . . . and numerous problems related to the building of the army which are left over from history'.[18] In short, the overriding financial fact in the development of the PLA throughout the Deng period has been inadequate funding.

This fact accounts for the entry of the PLA into the realm of trade and industry with the express purpose of earning money, and for the subsequent proliferation of its economic activities. Although these activities eventually produced undesirable offshoots, they have been encouraged by the leadership. Aside from the general rationale that military contributions to economic development also serve its own specific interests, the more mundane and relevant reason for this encouragement has been that money earned by the PLA is money that goes to making up the shortfall in military appropriations. A former director of the PLA's General Political Department put it this way in 1988:

> Our country's economic situation and the difficult tasks that it faces predetermine that it is impossible to expect a relatively big increase in outlays for national defense in the near future. One important measure for strengthening the building of the army is to vigorously expand production and operations so as to increase the army's ability to develop and perfect itself.[19]

The notion that the PLA should earn profits to support itself was supplemented by a second factor – the economic reforms of the Deng leadership. Military policy in China has always closely conformed to national policy. Therefore, without the new economic policy such a radical departure from the PLA's tradition would have been impossible. At the core of this policy was a rejection of Mao's anti-materialistic legacy and the launching of far-reaching reforms based on the concept that it was good for the nation if people got rich. And what was good for the people was good for the PLA. The reforms implemented in all sectors of the economy gave the PLA the sanction and the example to launch its own profit-making operations.

In the freewheeling climate created by the reforms, these operations grew rapidly, especially during the second half of the 1980s, as military appropriations decreased. Their growth was also undoubtedly fostered by an internal dynamism. Once the PLA crossed the barrier into the world of business, it was tempting to set up more and more money-making companies. Aside from new enterprises, many companies simply spawned subsidiaries.

A big push to PLA business ventures was apparently given by the upsurge of economic activity across the country that followed Deng's southern tour of early 1992. During this tour, Deng reportedly came down again, forcefully and unequivocally, on the side of reforms, including those containing capitalist elements, and blasted leftist opponents. The effect on the PLA was probably twofold. First, it weakened whatever opposition existed among senior commanders to commercial ventures. Second, it gave PLA units the green light to intensify their economic activities. After the tour, the Deng leadership enlisted the symbolic support of the PLA as an 'escort' of the reforms, a move that strengthened the military's identification with the market economy and reinforced the sanction to participate in it.

After a decade of economic involvement, the PLA has built up an empire that embraces a wide range of pursuits. However, the dimensions of this empire are not known. It is likely that what is visible to outside observers is only the tip of an iceberg the real dimensions of which are not even known to Chinese leaders. This is because, as already noted, PLA units may hide their economic activities from higher authorities, and military authorities may hide facets of the PLA's involvement from civilian leaders. For example, an official of the Xinxing corporation, which is controlled by the General Logistics Department, discovered upon visiting one of the conglomerate's plants that it had eight additional unreported businesses, including tourist shops and a guest house.[20] The most that can be done, therefore, is to draw up a sketch that is necessarily impressionistic and fragmentary.

What emerges from such a sketch is that PLA units are deeply involved in activities that cover every major sector of the economy:

agriculture and industry, the service sector and real estate, trade and finance. This involvement cannot be quantified: the figures published by the Chinese divulge very little and are often inconsistent. But some of the figures are indicative of its scope. By 1993 there were more than 10,000 production and industrial units run by the PLA,[21] although some military officers said unofficially that the number could have been double that.[22] This number obviously does not include enterprises affiliated with militia units, which are also under the supervision, partial at least, of the provincial military authorities, and whose inclusion would raise the number immensely.[23] According to PLA officials, about 1,000 PLA companies operate in Guangdong province, but some analysts estimate there are actually five times this number.[24]

In January 1992 it was reported that in the previous few years army-run enterprises had developed more than 2,500 products for civilian use.[25] In 1990, such enterprises were said to have produced 3.3 billion yuan worth of civilian products, which accounted for 55 per cent of their total output.[26] The importance of this production is clear: since the early 1990s, the income which it generated has covered about one-fifth of military expenditure deficits.[27]

A few random examples point to the variety of PLA activities. In Liaoning province, the Military District established more than 130 military-run enterprises since 1985, which produce food, petrochemical goods, and machinery, and are also involved in catering and service sectors. These enterprises employed some 5,000 staff members and the value of their output was about 730 million yuan, of which 116 million were profits, 21 million were paid in taxes, and 78 million went to the PLA.[28] In the Nanjing Military Region, the staff headquarters, the political department, and the logistics department established in recent years 26 enterprises, including factories, hotels, and guesthouses.[29] The Xinjiang troops reportedly were active in tourism and border trade, exporting products such as corn, flour, and rice to CIS countries.[30] Enterprises under the Chengdu Military Region capitalized on the region's special features and utilized folk recipes and herbs to produce goods such as tea, mineral products, and wine that were exported and earned foreign currency.[31] In Hainan the PLA ran gasoline stations, with soldiers pumping gas for civilians.[32] In some cases, the desire for economic gains ran wild. For example, a certain military medical school ran a total of 72 enterprises which engaged in production and business and in which all its departments and sections had a hand.[33] Sometimes the urge to make money drove PLA units to strange lengths. In Canton, for example, a TV station had a programme in which soldiers competed in simulated war games and viewers bet on the likely winners. And near

Tianjin, there was a PLA division which tourists could join as soldiers for a day.[34]

The air force and the navy are particularly suited and well placed to undertake large-scale projects. By 1993 the air force had contracted to build 28 large and medium civilian and military airports. It has eight airport construction units, staffed by skilled personnel and possessing sophisticated equipment, spread out over the country. These assets enabled it to outbid most competitors.[35] The air force has also gone into the civilian airline business. For example, the China United Airline Company, which is run by the air force, had opened 39 domestic routes by 1992 and has flown nearly 30,000 civilian sorties.[36] The Lantian (Blue Sky) trading company of the air force in 1985 launched its own commercial airline that competed with regional carriers for lucrative routes.[37] Moreover, the aviation maintenance industry under the Chinese air force provides a variety of services to some countries and also exports products for civilian use.[38]

The navy, for its part, established a shipping fleet for civilian needs, employing naval ships that used naval ports.[39] Units of the South China Fleet stationed in special economic zones established more than 460 construction companies in cooperation with foreign enterprises, state factories, and the localities.[40]

At the same time, the PLA continued its traditional agricultural production on army-run farms, with the produce going not only for its own consumption but for the market as well.[41] It also continued to participate in large-scale state construction projects, but with a difference: since early 1988 PLA units can be paid for their services.[42]

One sign that suggests the scope of PLA economic activities is the merging of enterprises into conglomerates. By 1993, twenty were set up.[43] The largest is the China Xinxing Corporation, which was founded in 1984 by the General Logistics Department as part of the effort to sell on the civilian market goods that were produced by its enterprises for the PLA. It controls more than 100 factories employing 10,000 workers.[44] Another conglomerate is the 999 Enterprise Group which combines all companies operating under the General Logistics Department in the Shenzhen special economic zone. It consists of 34 enterprises and has 1.6 billion yuan in fixed assets. In December 1992, it was reported that in the 10 months since its formation, the 999 Group had generated 157 million yuan in profits from activities in areas such as pharmaceuticals (999 is one of China's biggest producers), real estate, export–import, motor vehicles, electronics, food, clothing, trust investment, and stocks and securities.[45] The foreign trade companies of the Group have reportedly found markets in Russia, Sudan, Singapore,

Hong Kong, Taiwan, Egypt, and Qatar. In 1993 the volume of its foreign trade was said to be worth almost US$12 million.[46]

Also operating in Shenzhen is the PLA Changcheng (Great Wall) Industrial Conglomerate which belongs to the Guangzhou-based 42nd Group Army. It has some 90 enterprises which are operated by the Group Army's brigades and regiments.[47] The PLA has also established at least one Economic Development Zone. It is located in the Shantou special economic zone and is controlled by the Guangzhou Military Region. It used to be a farm run by local troops and is destined to become a large urban centre, which will have residential sections and whose activities will include commerce, industry, and tourism.[48]

A different activity, which is military in essence but has important economic aspects, is the sale of arms. The story of China's rise to a prominent position in the international arms market is well known. Suffice it to note that the volume of these sales has been enormous – it is estimated at US$21 billion for the years 1979–91.[49]

Most of the companies that sold weapons are affiliated with the military–industrial ministries and not with the PLA. However, the three PLA companies trading in weapons and equipment, each of which is controlled by one of its General departments, also took part in China's arms trade. The Xinxing Corporation under the General Logistics Department sold logistics equipment; the Kaili Corporation (Carrie Enterprises) under the General Political Department, dealt with communications equipment; and Polytechnologies, which is controlled by the General Staff Department through its Armaments Sub-Department, is one of China's main arms merchants.[50]

The connection between the arms trade and the PLA's economic involvement has two aspects. First, most of the profits made by these companies apparently go to PLA organizations that own them and are presumably used for current expenses and for investment in commercial ventures. Second, some of the weapons trading companies are branching out into other areas, such as property development[51] – thus creating new links between the PLA and the civilian economy.

In sum, the results of this inquiry into the economic involvement of the PLA are far from satisfactory. For one thing, the figures published by the Chinese are incomplete and not always reliable. There is, therefore, no way of measuring the dimensions of involvement in relation to the size of the PLA, and no way of determining how much of the PLA's total time and effort are devoted to economic activities. Second, the evidence of involvement is largely anecdotal and spotty. There is, therefore, no way of determining how representative the known activities are of the entire PLA. Despite the limitations, a firm conclusion can be drawn from the available evidence: numerous PLA units across the country are involved in a wide variety of profit-making activities to an extent that is bound to have an impact on their functions as military organizations. An attempt to assess this impact is the purpose of the following section.

THE EFFECTS OF INVOLVEMENT

If the dimensions of the military's economic involvement are uncertain, what is certain is that, along with its beneficial contributions to the finances of the PLA, this involvement has spawned negative consequences. Some of these can be gleaned from the Chinese media, but there is no doubt that what is revealed is extremely limited. The Chinese leadership is not interested in publicizing undesirable phenomena which point to a weakening of values that it wants to perpetuate or cast doubt on its ability to control the activities of the military. When it does so, it is for one of two reasons: either it deems such publicity useful for pedagogical purposes; or it has to publicize such phenomena in order to prohibit them. In any case, the yield from the media is necessarily meagre. Even so, it provides clues to what happened in the PLA in the wake of its economic pursuits.

What happened, first of all, was that once the traditional barriers to commercialism collapsed, PLA units rushed into the economic sphere in a manner that went far beyond the limits foreseen or desired by the leadership. This was probably inevitable, given the favourable environment created by a combination of several factors: the encouragement of the leadership; the spirit of rampant materialism that has characterized China in the era of reforms; the assets readily available to PLA units to take on economic undertakings; and the desire of officers and men to make up for past deprivations and to keep up with economic achievers in other sectors. The result has been a wide range of activities that are downright illegal.

The extent of these transgressions is evident, for example, in the 'ten nos' regulations, published in early 1989, which prohibited the PLA from engaging in a variety of activities such as: setting up businesses without permission, buying up goods illegally, jacking up prices, and engaging in profiteering; importing parts in order to assemble goods that were prohibited; producing and selling fake or shoddy products; using equipment without authorization; using military means of transportation and warehouses in order to carry out speculation, profiteering, and smuggling; hiring out or selling military vehicles, licences, bank accounts, and blank invoices; hiring out cultivated land and selling real estate without authorization; using officers and men on active service to run enterprises and engage in trade; providing labour services without authorization and ordering military personnel to engage in business ventures; and exploiting the positions of military men for business purposes.[52]

However, issuing regulations is one thing, complying with them, quite another. It was not that there was a serious breakdown of discipline in the PLA. It was just that the climate and conditions were conducive to circumventing such sweeping regulations, especially since the leadership seemingly made no serious effort to enforce them. Throughout the period of expanding PLA involvement in economic activities – the late 1980s and the early 1990s – the leadership apparently did not launch one major public campaign which was accompanied by specific measures aimed at stemming the undesirable consequences of the military's economic activities. What it did do was periodically to reiterate instructions aimed at eliminating illegal activities, but this only suggested that violations continued to occur.

Thus, for example, it was reported in July 1993 that the Central Military Commission had issued a circular drafted by the General Logistic Department that was designed to tighten PLA financial management. The main points of the circular decreed that the PLA units were not allowed to do the following: carry out financial ventures without permission; deposit PLA funds in individual accounts; speculate in foreign currency; develop grandiose projects, particularly guest houses, building, and holiday resorts; and speculate in real estate.[53]

Although no major campaigns against economic transgressions were launched in the PLA, the high command did take a number of measures against them. One was the establishment of accounting centres in all the Military Regions which are subordinate to the logistics departments of the Regions and to the General Logistics Department. All military units are supposed to deposit all their profits in these centres rather than in private accounts in civilian banks.[54] Another measure is the periodic inspection of units by the General Logistics Department. Thus, for example, it was reported in mid-1993 that the inspection of that year would centre on large PLA-run enterprises and would investigate the following aspects: evasion of state taxes; launching projects in violation of regulations; retaining, misappropriating, or delaying the delivery of profits; falsifying and concealing reports; diverting military budgets to profit-making operations; illegally speculating in real estate and foreign exchange; and purchasing goods which were prohibited or controlled.[55]

In an important effort to limit the effects of the PLA's economic involvement, the high command tried to separate military from economic activities in the armed forces. The principle, according to an article in the army paper, is that enterprise ownership should be separate from its management, that the organization setting up the enterprise should be separate from the one running it, that economic enterprises should be separate from military organizations, that enterprise personnel should be separate from military personnel on active service, and that the names of the enterprises should be separate from the code names of military units. Furthermore, enterprise managers should not be included in the military establishment, they should not wear army uniforms, and their pay should not come from the military budget.[56]

So much for the principle, but, according to a Chinese source, although it was first enunciated by the Central Military Commission in 1988 after an all-PLA meeting, the principle had not been put into practice five years later. Why? Because of the 'selfish departmentalist mentality' of officers. As a result, officers devoted too much energy to economic affairs, army units competed for materials and markets, they operated at low levels of efficiency, and they caused waste. This had damaged the image of the army.[57]

The efforts to uproot the offshoots of the PLA's ventures were obviously unsuccessful and by the early 1990s there were clear signs of concern. This was reflected in official pronouncements which, as usual, were short on details. An article in the army paper in December 1991 said that 'people . . . are . . . very worried about a few malpractices in the Army's production and economic activities'.[58] One of these 'people' was apparently Xu Xin, former deputy chief-of-staff-, who said in an interview that it was 'inappropriate for troops to set up companies and get involved in business activities'. While he did not rule out the sale of products from the PLA's farming and sideline activities, he emphasized that the military should not set up trading enterprises of a commercial nature.[59] Other articles in the army paper criticized 'money worship', which weakened 'selfless devotion',[60] and condemned ostentation and extravagance in personal relations that accompanied the growth of materialism.[61] Defence minister General Chi Haotian said that PLA members must oppose 'mamonism, hedonism, and extreme individualism', and 'resist corruption, never touch it'.[62]

But touch it they did. Cases of corruption are not likely to receive official publicity, but given the combination of the current money-making frenzy with the traditional culture, in which a certain amount of graft was tolerated, it may be assumed that they are considerable. This is indicated by occasional reports in the Hong Kong press – of 300 soldiers involved in a US$500,000 scandal in Liaoning province; of senior officers arrested for corruption and other illegal activities;[63] of several hundred army enterprises closed, and officers punished for corruption;[64] of the navy smuggling more than 2,000 cars and buses from Russia and South Korea, using a gunboat to thwart the customs police.[65]

The most significant sign of the PLA leadership's concern with the negative side of economic involvement was contained in an article co-authored by China's two top military commanders, Liu Huaqing and Zhang Zhen, for Army Day, 1993. The gravity of this concern was highlighted by the unprecedented joint authorship and the occasion for the publication of the article. Most of it is devoted to a glorification of the PLA's traditions of sacrifice, struggle, and discipline. However, the article cautions, it is these traditions that are threatened by the new economic situation. Officers and troops cannot avoid participating in the market economy, but this will bring new ideas and lifestyles which will 'give rise to money worship, hedonism, extreme individualism, and other erroneous ideas that will erode and buffet people's ideals and convictions'. Furthermore, the 'unhealthy tendencies and negative phenomena' that are prevalent in the society are also infiltrating the Armed Forces through various channels, jeopardizing military construction. The authors go on to sound a warning: there are many historic examples, in China and elsewhere, which show that in a peaceful environment the military 'became pleasure seekers, lost their combat readiness and were eventually defeated by peace, as well as by themselves'. The PLA, they emphasize, must be aware that 'things develop and change according to certain conditions'. In order to withstand the new dangers, it must preserve and develop its revolutionary traditions.[66]

This the leadership has tried to do by exhorting the PLA to follow the examples of Lei Feng – the legendary soldier who was motivated by boundless self-sacrifice – as well as of the 'Good Eighth Company of Nanking Road' and its successor, the 'Good Eighth Company of Gulungyu' – model units which had become a beacon to the PLA because their revolutionary qualities had enabled them to overcome the temptations of a bourgeois environment.[67]

The campaigns to emulate these and other models had indeed been instrumental in reviving revolutionary traditions in the PLA, but they had been conducted in an entirely different setting. In the early 1960s, Maoist ideals were intensely propagated in the society at large; China was virtually cut off from outside developments and the campaigns in the PLA were forceful and sustained. In the 1990s, however, not only can the PLA not remain insulated from the commercialism that dominates a society in transition to capitalism, but its own economic activities foster these traits within its ranks. In this milieu, the exhortations to reassert revolutionary qualities – anaemic to begin with – are surely met with indifference, if not worse, by officers preoccupied with weaving financial deals. A serious effort to cleanse the PLA of 'unhealthy tendencies' plainly requires much stronger measures – but these have not been forthcoming.

This is definitely not because the Chinese military leadership is unaware of what measures are required. If it needed reminding, a February 1994 article in a Hong Kong paper referred to an analysis produced by a group of scholars in the Chinese Academy of Sciences which clearly laid out what has to be done in order to extricate the army from its dangerous economic involvement. The analysis seems authentic, but even if it were not produced by this organization, its proposals are eminently logical.

After warning that the PLA's economic activities may lead to its 'commercialization' and 'localization', the analysis points out that in the few countries where the military are deeply involved in business, corruption is notorious and the armed forces developed special economic interests which are independent of state interests. To avoid this, it suggests several strong remedies. First, the PLA should be prohibited from engaging in commercial activities of any kind and must not accept appropriations from local governments. Second, the military budget should be increased substantially and all PLA expenses should be covered by the government. Third, some of the items required by the troops, such as food, clothing and medicines, need not be produced by the army but should be obtained from civilian suppliers. Fourth, military industrial enterprises should be separated from the regular military establishment, with the exception of a few key and confidential enterprises. Fifth, the armed forces should not be used as a low-cost labour force.[68]

Why then has the leadership not taken, at least not in a determined and sustained way, at least some steps to curb the PLA's economic activities? Several possibilities come to mind. One is that they are simply unable to enforce dictates, but this possibility can be ruled out because it implies that the high command has virtually lost control over the armed forces and that is certainly untrue. More likely is a combination of two other factors. The first is that the deleterious effects of the PLA's economic entanglement are not yet serious enough to warrant immediate and major countermeasures. The process which may erode the PLA's military capabilities and professional qualities is gradual and largely invisible, and its full impact will be felt only in the future. For this reason, China's military leaders may have concluded that at this stage it is enough to issue warnings and to initiate educational measures, but not to take more serious steps.

This leads to the second factor. More serious steps would mean nothing less than a large-scale reduction of the army's economic activities, and the high command is obviously not ready to do this. Such a reduction would probably cause a crisis unless PLA units are compensated by a huge increase of funds, which are not available to the military leadership. More important, the PLA's economic activities fulfil enormously important positive functions: they ease financial pressures on the military leadership; they provide it with funds that give it considerable freedom of action; and they upgrade the standard of living and the military capabilities of PLA units. In short, there are palpably powerful reasons against cutting down on economic activities, the benefits of which are tangible, whereas the dangers which they embody are largely potential.

What, then, are these dangers? The answer is speculative because, as has been pointed out, the connection between economic involvement and its effects is not necessarily concrete and conspicuous but has to be inferred from the army's activities, the data about which are sparse. However, from what is known about these activities, it is feasible to consider their possible impact.

First, there is the impact on combat capability, which has two components. One is technological, and if there is any connection between weapons procurement and economic involvement, it is entirely positive. Since the start of the 1990s the PLA has stepped up weapons acquisition and development, as indicated most cogently by the purchase of SU-27 aircraft and advanced surface to air missiles (SAM) from Russia. One reason for this has doubtless been the Gulf War; another, the desire to compensate the PLA for years of limited appropriations; a third, the ready availability of sophisticated weapons. However, new weapons are expensive, and it is conceivable that the income generated by the PLA's economic activities accounts in part for the increased allocations to weapons development.

The other component is human – the training and all the other routine tasks that go into maintaining and raising the standards and skills of the troops. Have these tasks been adversely affected by the PLA's economic pursuits? Judging by the pronouncements of senior military leaders and articles in the army paper, the answer is a resounding no. Upgrading the military levels of PLA soldiers has been a top priority objective of the military leadership throughout the post-Mao period of military modernization, and this has not changed as the army expanded its non-military activities. If anything, the importance of constantly preparing officers and men for combat has been reasserted over and over again by China's top commanders.[69]

However, whether this objective been attained in the far-flung units of the PLA is an open question. When opportunities for profit-making abound in the present, the temptation to cut corners on military tasks is likely to be too great to overcome, especially since the damage will become apparent, if at all, only in the future. The main victim of this temptation is probably military training. If the waking hours of a peacetime army are supposed to be devoted to training and other essential activities, and if troops are diverted to other occupations, this must inevitably be at the expense of military missions. Therefore, large-scale economic involvement cannot but have a corrosive effect on the military posture of the PLA.

How corrosive depends on several factors which cannot be determined with any certainty. One, already mentioned, is the extent of the PLA's economic involvement. Another is the identity of units involved in economic undertakings. If most of these units belong to rear service echelons, the effect on combat capability is obviously less severe than it would be if they were first-line divisions. While no reliable data are available to ascertain this, the impression is that combat divisions which are designated as part of China's rapid deployment force are minimally involved, if at all, in economic activities. In fact, they may benefit from the income gained by other, less combat-oriented, units. A third factor pertains to the number of soldiers assigned to running, and working, in the numerous military enterprises. If an enterprise is owned and managed by a military unit but employs civilians (where possible, probably relatives of soldiers) the damage to the military capability of the unit will plainly be less than if the entire unit is employed on, say, a construction site.

However, even without precise data on these points, there is little doubt that a large number of officers in many PLA units have found a new and additional, if not exclusive, vocation. And this vocation is a threat to military professionalism – to its basic elements of specialization, corporate separateness, discipline, and subordination to political control. The reasons are plain. Specialization, it need hardly be emphasized, requires complete dedication to the ongoing acquisition of new knowledge and its absorption through sustained training. This is particularly true of the PLA in the 1990s, as it procures increasingly complex weapons and adopts new war fighting doctrines. However, such dedication is not likely to be demonstrated by officers who are busy managing enterprises or conducting financial affairs in a tense competitive setting. In the clash between the battlefield and the boardroom, it is military needs that surely suffer.

What also surely suffers is the corporate spirit that bonds officers with a sense of belonging to a brotherhood distinguished by a singular calling and a unique way of life. The preoccupation of officers with economic pursuits cannot but weaken the ties that bond. For these pursuits remove officers from a strict military setting that fosters the growth of corporate sentiments. They also enhance identification with a particular unit for reasons extraneous to the military calling rather than with the broader military entity.

The result may be the rise of a new type of factionalism, which is fuelled by economic particularism and the related rivalries that it generates between PLA units. Such factionalism could open up new divisions within the PLA and obstruct cooperation between units that should be based solely on military considerations. More ominously, it could impel commanders to give precedence to parochial economic interests over wider army needs. And it could result in the formation of informal alliances between PLA units with common economic interests aimed at bringing pressure on upper echelons or thwarting their instructions.

The economic imperative may also endanger military discipline. In a possible conflict between the pull of profits and obedience, profits may prevail. Officers enjoying new status and power that derive from their economic success are more liable to take unprecedented liberties in disregarding orders that jeopardize lucrative economic operations. And officers who have honed the entrepreneurial skills essential to making it in the business world are more liable to buck at carrying out orders which conflict with their commercial needs. That this has already happened was demonstrated by the obvious noncompliance of military units with regulations that were aimed at curbing or abolishing a wide range of their economic activities. Such infractions are apparently not considered serious enough by the leadership to warrant effective countermeasures, probably because they occur outside the mainstream of military affairs, and because the leadership is reluctant to put the brakes on profit-making activities. However, discipline is indivisible, and if violations are tolerated long enough, erosion might set in.

If this happens, the implications for the internal functioning of the PLA and for political control of the military may be farreaching. This is not only because an erosion of discipline will weaken the sinews which, above all else, hold a military organization together. It is also because the pivotal importance of discipline becomes all the more crucial as China enters an uncertain period of leadership transition.

Symbolized by the passing of Deng Xiaoping, this transition marks the final disappearance of revolutionary leaders in the Communist Party and the PLA – leaders whose ability to command compliance from military commanders did not derive only from their position at the pinnacle of the military hierarchy, but also, and at times much more, from their personal stature and influence. Their successors, however, do not have such stature. Therefore, the future ability of party leaders to control the military, and of PLA leaders to control the military command hierarchy, will rest, much more than at any time in the past, on the bedrock of organizational discipline – the unquestioned execution of orders and the routine adherence to regulations. And it is precisely this bedrock that is in danger of being undermined by the PLA's economic involvement.

The immediate and most serious danger to the PLA derives from corruption. Rampant in the increasingly materialistic Chinese society, corruption has also spread in the PLA to what is apparently a considerable extent. It may take many forms – acceptance of bribes, disobedience for financial gain, participation in illegal activities, to name but a few – but whatever the form, its effects are devastating. Above all, it demolishes the ethic which lies at the heart of professionalism – the ideal of duty and service to country that impels officers to choose the military not as a profitable career, but as a calling without regard for personal gain.

No army can function properly if corruption in it is widespread. Until the 1980s, this had never been a serious problem in the PLA. The absence of tempting opportunities combined with stringent organizational and ideological controls proved to be an effective barrier against corruption. The exception, perhaps, was the Cultural Revolution period, although even then, serious corruption, to the extent that it existed, was probably not for financial benefits. However, the extensive economic involvement of the PLA provided unprecedented opportunities for corrupt practices. At the same time, ironically, the strengthening of professionalism in the PLA and the end of ideology in society weakened the political apparatus in the armed forces and greatly reduced the effectiveness of ideological education among the troops. The result has been a fertile breeding ground for corruption.

If the trends noted in this chapter remain unchecked, the effect on civil–military relations may be far-reaching. As a starting point, they may lead to the formation of a PLA that has accumulated vast economic power but, as a result, has become weak, if not worse, in discipline and cohesion. The implications that could arise from this situation are ominous. One is unprecedented autonomy of the PLA high command from the civilian leadership due to its reduced dependence on financial allocations from the government. The second is a refusal of the military leadership to follow government orders which impinge on its economic interests. The third is the refusal of lower PLA echelons to follow government orders, even though these have been transmitted by the military leadership. The fourth is the creation of coalitions between PLA groups with special economic interests – such as, for example, senior members of the General Logistics Department – and civilian politicians with similar interests. The fifth is the formation of local alliances based on common business benefits between PLA commanders at the regional level and below and their counterparts in party and government organs. The sixth is the blurring of the PLA's distinct institutional identity, which is anchored in overriding professional interests, and the emergence of a PLA that in reality is made up of 'one army, two systems' – one military and one economic.

The Chinese leadership is, of course, aware of the dangers to the PLA's professional military posture that are inherent in its manifold commercial activities. At the beginning of 1995, it was reported that the campaign to reduce the number of military-run enterprises had been successful and that active-service personnel had withdrawn from business ventures.[70] However, given the extent of the PLA's economic involvement, and considering previous gaps between statements and reality, the final verdict on this critical issue still cannot be given.

NOTES

1 Beijing, *Xinhua*, Domestic Service, 3 July 1993; *FBIS*, 15 July 1993, p. 23.
2 One source puts the figure at 66 per cent, another at 76. See Beijing *Xinhua*, 2 February 1993, 3 February 1993, p. 28; and *RenMin RiBao*, 27 July 1992; *FBIS*, 13 August 1992, p. 27.
3 Mel Gurtov, 'Swords into Market Shares: China's Conversion of Military Industry to Civilian Production', *The China Quarterly*, June 1993, no. 134, pp. 214–16; and Yitzhak Shichor, *Military to Civilian Conversion in China: From the 1980s to the 1990s*, Peace Research Centre, The Australian National University, Research School of Pacific Studies, Working Paper no. 142, December 1993, pp. 3–10.
4 Gurtov, pp. 216–20.
5 Beijing *Xinhua*, Domestic Service, 31 June 1990; *FBIS*, 15 February 1990, p. 76.
6 Ibid., 3 July 1993; *FBIS*, 15 July 1993, pp. 23–4.
7 Ibid., 6 December 1993; *FBIS*, 15 December 1993, p. 27.
8 See note 5.
9 Ibid.
10 See note 7.
11 *Far Eastern Economic Review*, 14 October 1993, pp. 70–1.
12 *Cheng Ming*, 14 October 1993, pp. 70–1.
13 Gurtov, p. 222.

14 Gurtov, p. 229.
15 Tai Ming Cheung, 'Profits over Professionalism: The People's Liberation Army's Economic Activities and its Impact on Military Unity', paper presented to the conference on the 'Security Dimensions of Chinese Regionalism', sponsored by the International Institute for Strategic Studies and the Chinese Council of Advanced Policy Studies, Hong Kong, 25–7 June 1993, p. 4.
16 *Jiefangjun Bao*, 12 December, 1991; *FBIS*, 13 January, 1992, p. 40.
17 Tai Ming Cheung, pp. 4–5.
18 *Jiefangjun Bao*, 31 March, 1987; *FBIS*, 10 April 1987, p. 25; and *Renmin Ribao*, 8 January 1989; *FBIS*, 12 January 1989, p. 25.
19 Beijing, *Xinhua*, Domestic Service, 29 April 1988; *FBIS*, 3 May 1988, p. 30.
20 *Far Eastern Economic Review*, 24 October 1993, p. 65.
21 Beijing *Xinhua*, Domestic Service, 29 December 1992; *FBIS*, 5 January 93, p. 31.
22 *Far Eastern Economic Review*, 14 October 1993, p. 64.
23 In the Nanjing MR alone there are said to be 12,000 militia-affiliated enterprises. Beijing Central Television Program One Network, 28 May 1993; *FBIS*, 7 June 1993, p. 31.
24 *Far Eastern Economic Review*, 14 October 1993, p. 65.
25 *Jiefangjun Bao*, 12 December 1991; *FBIS*, 13 January 1992, p. 40.
26 *Jiefangjun Bao*, 16 December 1990; in TaiMing Cheung, p. 6.
27 *Jiefangjun Bao*, 23 July 1992; 14 August 1992, in Tai Ming Cheung, p. 6.
28 *Liaoning Ribao*, 4 April 1993; *FBIS*, 27 April 1993, p. 31.
29 Note 25, p. 42.
30 *Hong Kong Zhongguo Tongxunshe*, 16 May 1993; *FBIS*, 18 May 1993, p. 33.
31 *Jiefangjun Bao*, 25 October 1991; *FBIS*, 15 November 1991, p. 45.
32 *The China Business Review*, Sept–Oct, 1989, p. 30.
33 Note 25.
34 *Far Eastern Economic Review*, 14 October 1993, p. 65.
35 *Jiefangjun Bao*, 4 October 1993; *FBIS*, 14 October 1993, p. 20.
36 Beijing *Xinhua*, Domestic Service, 7 July 1992; *FBIS*, 9 July 1992, p. 29.
37 *The China Business Review*, op. cit.
38 Beijing, *Xinhua* in English, 4 February 1993; *FBIS*, 5 February 1993, p. 12.
39 *Jiefangjun Bao*, 2 June 1992; *FBIS*, 23 June 1992, in Tai Ming Cheung, p. 12.
40 Beijing, *Xinhua*, Domestic Service, 9 April 1993; *FBIS*, 20 April 1993, p. 22.
41 Note 25, p. 40.
42 Beijing, *Xinhua* in English, 30 July 1988; *FBIS*, 1 August 1988, pp. 46–7.
43 *Jiefangjun Bao*, 18 February 1993; *FBIS*, 2 March 1993, p. 27.
44 *China Daily*, 14 August 1986, 22 October 1992; in Tai Ming Cheung, pp. 12–13.
45 *Zhongguo Xinwen She*, 29 November 1992; *FBIS*, 9 December 1992, p. 33.
46 *Jiefangjun Bao*, 10 June 1993; *FBIS*, 21 September 1993, p. 43.
47 *Jiefangjun Bao*, 1 December 1988, 26 January 1992; in Tai Ming Cheung, p. 13.

48 Beijing, *Xinhua* in English, 12 February 1993; *FBIS*, 12 February 1993, p. 17.
49 Schichor, p. 23.
50 Eric Hayer, 'China's Arms Merchants: Profits in Command', *The China Quarterly*, December 1992, p. 1113.
51 For an excellent report on Polytechnologies going Civilian, see the article by Tai Ming Cheung in *Far Eastern Economic Review*, 14 October 1993, p. 68.
52 *Hong Kong Zhungguo Tongxun She*, 25 April 1989; *FBIS*, 2 May 1989, p. 116.
53 *Beijing, Xinhua Hong Kong Service*, 21 July 1993; *FBIS*, 26 July 1993, pp. 32–3.
54 Tai Ming Cheung, p. 7.
55 Tai Ming Cheung, p. 4; Beijing, *Xinhua*, Domestic Service, 28 August, 1993; *FBIS*, 1 September, 1993, p. 26.
56 *Jiefangjun Bao*, 13 December 1991; *FBIS*, 13 January 1992, p. 41.
57 Ibid.
58 *Jiefangjun Bao*, 12 December 1991; *FBIS*, 13 January 1992, p. 41.
59 *Zhongguo Xinwen She*, 1 March 1993; *FBIS*, 9 March 1993, p. 37.
60 *Jiefangjun Bao*, 9 July 1993; *FBIS*, 30 July 1993, p. 29.
61 *Jiefangjun Bao*, 14 July 1993; *FBIS*, 5 August 1993, pp. 29–30.
62 *Jiefangjun Bao*, 6 September 1993; *FBIS*, 17 September 1993, pp. 36–7.
63 *Far Eastern Economic Review*, 12 August 1993.
64 *Hong Kong Standard*, 3 April 1990; *FBIS*, 3 April 1990.
65 *Hong Kong Cheng Ming*, 1 September 1993; *FBIS*, 10 September 1993, pp. 35–6.
66 *Renmin Ribao*, 26 July 1993; *FBIS*, 27 July 1993, pp. 22–5.
67 See, for example, *Shanghai Jiefang Ribao*, 26 April 1993, pp. 21–4; Beijing, *Xinhua*, Domestic Service, 8 March 1993; *FBIS*, 16 March 1993, p. 47; *Jiefangjun Bao*, editorial, 5 March 1993; *FBIS*, 12 March 1993, pp. 18–19; *Jiefangjun Bao*, 4 March 1993; *FBIS*, 11 March 1993, pp. 27–9.
68 *Hong Kong Lien Ho Pao*, 17 February 1994; *FBIS*, 25 February 1994, pp. 29–30.
69 See, for example, statement by Chief-of-Staff Zhang Wannian, Beijing, *Xinhua*, Domestic Service, 3 December 1993; *FBIS*, 8 December 1993, p. 24; article by Liu Huaqing, first vice-chairman of the Central Military Commission, *Jiefangjun Bao*, 6 August 1993; *FBIS*, 18 August 1993, pp. 15–22.
70 Jietangĵun Bao, 10 February 1995, in Taī Ming Cheung, *China's Entrepreneurial Army: The Structure, Activities and Economic Returns of the Military Business Complex*, prepared for the 6th Conference on the PLA, West Virginia, 9–11 June, 1995, p. 5.

3 Corruption in the PLA

David S. G. Goodman

Campaigns against corruption – moral, political, as well as economic – are nothing new in the People's Republic of China (PRC). However, it would appear that the leadership of both the Chinese Communist Party (CCP) and the People's Liberation Army (PLA) developed an increased concern with corruption during 1993. In December 1992 the CCP's Central Commission for the Inspection of Discipline (CCID) held a conference in Shenzhen to examine 'Theory and Practice in the Struggle against Corruption',[1] which set the agenda for the following year. Zhang Siqing, Procurator-General of the Supreme People's Procurate reported on the results of the campaign against corruption during the first half of 1993;[2] and at the end of August Jiang Zemin delivered a major speech on the subject at the second plenum of the CCID of the Fourteenth Central Committee of the CCP.[3] Specifically within the PLA the various concerns about corruption were echoed by Liu Huaqing and Zhang Zhen, Vice Chairmen of the Central Military Commission, in July when they emphasized that 'we military comrades do not live in a vacuum' and were not immune from 'the various harmful trends and negative tendencies that have found their ways into and endanger our army'.[4]

In its analysis the CCP has emphasized the dysfunctionality of corruption and isolated four causes: the penetration of Chinese society by Western bourgeois values; the continuation of traditional Chinese autocratic values; the social and psychological dislocations of rapid economic transformation; and the low quality of cadres. Corruption is seen very much as moral decay, which is said to manifest itself in many different forms. For example, cadres of the party and state administration are regarded as having abused the privileges of their positions in return for personal gain, and to have traded political power for economic wealth. Departments concerned with personnel, financial and economic management have been particularly singled out for criticism.

Personnel from judicial and law enforcement bodies are criticized for having resorted to blackmail, and seriously bending the law for the benefit of their relatives and friends.

However, the identification and analysis of corruption in the PRC is by no means straightforward. Corruption is an essentially contested concept in the contemporary PRC, even though there are some activities and behaviours that may be universally agreed to be corrupt. Analysis is not aided by both the CCP's equation of moral, political, and economic corruption, and at the same time its inherently ambiguous and contradictory policies in some areas. One particularly pertinent example is the explicit policy that encourages military units to engage in economic activities while at the same time, as part of the anti-corruption campaign, cadres of the party–state system in general are discouraged from engaging in entrepreneurial activities. Moreover, in its recent past the CCP has frequently used the designation of corruption as a political weapon and demonstrated its facility for branding as corrupt behaviour that it has ex post facto decided to criticize.[5]

Scott's definition of corruption as 'behaviour which deviates from the formal duties of a public role because of private-regarding (personal, close family, personal clique) wealth or status gains, or violates rules against the exercise of certain types of private-regarding influence'[6] is clearly only of limited applicability in the PRC of the 1990s. Though that definition attempts to address and transcend the requirement to acknowledge relativism, often lost in cross-cultural comparison, it nonetheless overemphasizes the importance of the individual and the personal. For example, there is considerable evidence that cadres and entrepreneurs engage in apparently corrupt behaviour for collective, and not just personal, ends. It may be necessary to offer some kind of additional inducement to a supplier, in effect to bribe, in order to ensure production even when both supplier and producer are collective entities within the party–state system. Corruption in China may stretch from actions officially regarded as politically undesirable through morally unacceptable behaviour, to illegal actions, as well as the private use of public benefits.

At the same time it is important to identify the definer and source of perspective on corruption within the PRC. Some actions may at some times be officially regarded as corrupt by the CCP or even the procuratorate, but equally not at others or under all circumstances. For example, the taking of fees for services rendered by cadres is currently regarded as an illegitimate procedure by the central party and government. However, at the more local level a certain degree – which will vary with locality and activity – of fee-taking is not regarded by the population as corruption, or at least is regarded as a relatively acceptable procedure that does not offend local mores: a standard of permissible corruption.[7]

Moreover, despite CCP concerns it is also far from clear that corruption is totally dysfunctional. In general there is a considerable literature that emphasizes the extent to which corruption may also be functional in economically developing nations and political systems in transformation. Although it is clearly neither a comprehensive description nor a total explanation, corruption assists in the redistribution of public goods as societies become more open; provides moral and prudential guidance in the absence of other more formal regulation; enables administrative change to occur despite resistance; and ensures sustained economic output and administrative capacity. The free-rider problem may be greater in scale and extent than otherwise, nonetheless, corruption ensures that systems undergoing rapid transformation are able to adjust and continue in operation.[8] This would certainly appear to be the case to some extent with the emergence of corruption in and related to the PLA since the start of the reform era and particularly during the 1990s. Although the CCP and indeed the leadership of the PLA itself has concentrated on the debilitating effects of corruption on morale within the PLA, the PLA's general effectiveness, and on the PRC's defence policy,[9] corruption also enables the PLA to adapt to its new reform environment and objectives.

There are two major consequences of this perspective on corruption in the PLA. The first is that it emphasizes the inherently transitory nature of such behaviour. The second is that it also emphasizes the identity of the PLA with and within the party–state system. It is sometimes suggested that the PLA has a different organizational psychology to other parts of the party–state system, and may therefore come to play a somewhat special role in the process of regime transformation. Though this may indeed prove to be the case, the evidence of known or alleged examples of corruption within or related to the PLA in recent years would seem to suggest otherwise. Cases of bribery, fraud, and embezzlement involving the PLA (institutionally and associationally) are not structurally different to examples of corruption elsewhere in the party–state. Corruption in the PLA may involve its personnel, units, and enterprises, but for the most part there is little real military angle to such activities. The one corrupt practice which may justifiably be regarded as characteristic of the PLA more than any other part of the party–state system is smuggling. Units, groups and individuals have been able to manipulate the PLA's central role in domestic public security and its role as guardian of national borders to develop a smuggling industry in various ways.

THE CORRUPTION IMPERATIVE

China's reform era and the transition from a command economy to a market economy have presented considerable opportunities for corruption. In the three years from 1990 to 1992, discipline inspection and supervision organs placed on file and investigated 707,000 cases of violation of discipline; completed the investigation of some 650,000 of them; and punished more than 600,000 CCP members and cadres, including 16,005 cadres at and above the county level.[10] During the ten years from 1983 to 1992, a total of 2,247,900 cases of violation of discipline and the law were placed on file for investigation and prosecution. More than 1.8 million CCP members and cadres were punished or tried by courts for corruption or discipline violation, of which 745 were cadres at the provincial level, 7,890 at the prefectural level, and 56,474 at the county level. More than 188,100 members of the CCP in total were expelled.[11] The New China News Agency reported that procuratorates nationwide handled a record 50,000 cases of economic crime in 1993. Nearly half of them were instances of bribery or embezzlement, each involving more than 10,000 yuan RMB (Renminbi). Many culprits were officials of the CCP and state administration, including one vice minister, 46 departmental and 848 county level officials.[12]

The amount of funds involved in some cases has been sizeable by any standard. In the first half of 1993, there were 95 reported cases of corruption, bribery and embezzlement involving over one million yuan each.[13] The largest instance of embezzlement involved 30.44 million yuan and a group led by Xue Genhe, head of the accounting group in the Dongfeng Branch of the Haikou Industrial and Commercial Bank, Hainan Province. The largest example of bribery involved 7 million yuan – in a number of different bribes – accepted by Zeng Lihua, deputy manager of the Shenzhen Engineering Consultative Company.[14]

The CCP attributes such examples of corruption to moral decay – 'the pernicious influences of feudalism' and 'the rotten ideas of capitalism' – rather than to the structural imperatives of transformation.[15] However, it is quite clear that the latter are of considerable importance: they not only provide the opportunities for corruption, they also provide the stimuli. Inflation is the most obvious of these, for it has an immediate impact on cadres and officials on relatively fixed salaries. On the other hand inflation is fundamentally epiphenomenal: there is a general revolution of rising expectations where unmet demand is not restricted to questions of wealth or money. As the economy restructures itself organizational relationships and procedures must also change. Corruption may clearly result where supervision is weak, but it may also occur where regulatory frameworks have not kept pace with other changes, and policy direction becomes either contradictory or ambiguous. Thus in the current reform era the existence of a planned economic sector alongside a market-driven economic sector has allowed entrepreneurs (in both sectors) to gain maximum advantage from cross-sectoral activities. Such practices are further encouraged by the emphasis placed by the CCP on 'the development of productive forces' and economic growth: performance is no longer measured by political virtue but by economic output.

The relative decline of the state sector and the growth of the collective and private economies, decentralization, the introduction of market forces, and policies designed to encourage trade and investment within the international economy have all occurred relatively quickly and with inadequate infrastructure and regulation in the economy as a whole. With few exceptions the PLA's historically privileged position has not been preserved in this economic restructuring. It too has had to become more business oriented. The previously highly centralized military has been broken up into smaller parts.[16] PLA unit budgets have even been reduced specifically to encourage tighter economic management and the development of entrepreneurial activities.[17]

However, the PLA has retained certain privileges not available to the rest of the party–state system. One is its ability to shelter behind an appeal to the protection of national security when its own activities or procedures are under investigation. Information about corruption in the PLA is restricted and is classified as a military secret. As of 1 January 1993 responsibility for all cases of potential corruption and all economic disputes within the PLA were brought under the sole jurisdiction of military tribunals, and subject to military secrecy.[18]

Necessarily, the PLA has established its own regulatory framework in its attempt to combat corruption. Units that engage in military business are not permitted to create holding companies or any form of subordinated accounting unit that might (conceivably) siphon off funds. Equipment still in service cannot be used for business activities. Military units that engage in commercial or business activities are not given profit targets.[19] But in reality, practice of these sorts of activity is a common sight and not necessarily illegal.

However, economic corruption within the PLA probably cannot be regulated out of existence, not least because of rapidly increasing inequalities in personal income between service personnel, on the one hand, and the civilian new rich, on the other. It is now widely accepted that, in Gurtov's words, 'Inflation has lowered the day-to-day living standards of ordinary soldiers and made a military career very unattractive. PLA recruitment is therefore well down'.[20] By comparison with their peer groups PLA officers as soldiers can readily see how far they are falling behind 'keeping up with the Wangs'. Realizing the impact of psychological imbalance on military morale, one commentator in Liberation Army Daily – the PLA newspaper – desperately urged service personnel not to compare their standards of living with those of workers, cadres, and entrepreneurs in China's prosperous regions, but rather with peasants and civilian cadres in the backward regions.[21]

Corruption is also hardly avoidable when PLA officers and soldiers are under considerable pressure to cultivate personal relations with their superiors, local power holders, and business people in order to survive in their current positions and after demobilization. Since the start of the reform era in the late 1970s, massive cuts in military personnel have gone hand in hand with the streamlining of staff and workers in state organizations and enterprises. Except for an exceptionally favoured very few, there are no longer opportunities available for officers to be transferred to equivalent positions in local governments, or for soldiers to find jobs in state enterprises. In order to lay down the foundations for a good career after being discharged from the PLA, it is often the case that officers and soldiers find it necessary either to offer bribes or to involve themselves in private business, and sometimes even both.[22]

Under such circumstances it would be more surprising if some of China's service personnel did not find personal corruption irresistible. 'In the wake of economic liberalization and the development of military production, as Yang Baibing, then the Director of the PLA General Political Department and the Secretary-General of the Central Military Committee commented:

> economic crimes have also increased: the presenting and taking of bribes, embezzlement and stealing. With the deepening of reform and the readjustment of social interests, some party members and cadres are unable to maintain a proper balance in the relationship between the party organisation and the individual, and some even avail themselves of loopholes in the reform and exploit the advantage to reap profit.[23]

Certainly, some members of the PLA have attempted, as have some of their civilian counterparts, to commercialize their power in the party–state and seek private gain at public expense. However, various corrupt practices – including bribery, fraud, and embezzlement, as well as smuggling – occur for collective as well as private gain: indeed often the two are inseparable. A military unit may need to bribe to ensure its supplies. Production units within the PLA have to compete in the imperfectly marketized economy alongside civilian enterprises.

BRIBERY

A market economy is no more corrupt necessarily than a command economy. Nonetheless, it is clear that the introduction of market reforms since the early 1980s has resulted in a wide range of new business practices. The most common of these new business practices to be critically publicized include accepting an allowance or sale commission in exchange for issuing a contract on a project to a specific contractor; selling state sector products to buyers outside the state sector at a lower than market price; and for buying goods from the non-state sector at a higher than market clearing price. Examples of all these practices, which under certain circumstances may be regarded as corrupt, are readily available.

A PLA officer who was in charge of building materials for a division in the Jinan Military Region was disciplined at the beginning of 1987 for accepting a payment of 800 yuan from a private construction company.[24] In August and September 1988, a salesman from a company offered the Director of the Logistics Department in the Jiangsu Military District a bribe of 50,000 yuan as part of a purchase of 300 colour TV sets and some other goods from the company. A manager of a factory tried to bribe a section head, responsible for petrol allocations in the Logistics Department of the Nanjing Military Region, with 1,000 yuan for a project designed to involve 40,000 yuan of petrol. Two salesmen from a condenser factory offered the Director of the Communications Section of a PLA unit in the Nanjing Military Region a bribe of 2,000 yuan in exchange for goods ordered from the factory.[25]

In August 1993 the head of a construction company tried to bribe a division commander in the Xinjiang Military District to lower the quality target of a construction project on which his company was working.[26] As in much else, Guangdong Province has adapted more fully to the transformation to market forces. There service personnel are not only allowed, they are encouraged to take an allowance or sales commission when buying or selling. However, they are then required to pass any such payment on to their unit, though they will receive a certain proportion back as an award or bonus.[27]

Bribery is now reported to be fairly common even within military camps, particularly in the contexts of recruiting new party members, choosing candidates for military schools, deciding on awards, promotion, and transferring service personnel to civilian work.[28] Many soldiers are said to depend on their families for money to bribe officers for entry into the CCP, merit certificates, and transfer to a good civilian job after demobilization.[29] A soldier in Beijing was reported to have become a thief simply because he could not offer to bribe his officers enough to ensure promotion to the position of a specialized noncommissioned officer.[30]

FRAUD

At the same time, the PLA's preferential rights and special treatments granted by the state, particularly the access to scarce commodities and raw and other materials at lower prices, and the reduction and remittance of tax, can prove extremely lucrative. These preferential rights enable an officer or a military unit to make money simply by charging fees for using an official seal or a military registration form.[31] The most obvious of these and a practice that is fairly general around China is the sale of military registration plates, particularly for personal automobiles. Cars with military registration plates are given preferential treatment on China's crowded roads, and in the early 1990s were a major status symbol for China's 'new rich'. Commodity supply and purchase is another fairly obvious area where military resources are commercializable. In November 1988, for example, several senior retired officers of the Jiujiang First Retired Officers Sanatorium of the Jiangxi Province Military District were indicted for speculating in steel and timber supplies.[32]

The power and standing of the PLA is such that it is a suitable model for emulation in attempted fraud. In April 1993 a relatively large-scale case of apparently fraudulent military enterprises was exposed by the Shanxi Province Military Command and Department of Public Security. Though at first sight this case was presented as an example of a fraudulent military enterprise, it is not clear whether the fraud lay with the enterprise pretending to be associated with the PLA, or in the PLA engaging in such entrepreneurial activities in the first place.

Four gangs who were said to have falsely obtained more than 10 million yuan from PLA units, various enterprises and individuals through disguising themselves as service personnel and representatives of army-run enterprises were arrested. Their leaders – 'Major General' Feng Zhengzhang, 'Senior Colonels' Guo Rongxi and Wang Minglong, and 'Colonel' Li Tianyou (their ranks were reported as illusory) – were among those arrested. Several enterprises were closed down and categorized as 'fraudulent military enterprises', including the 'Beijing Military Region Logistics Department Military Supplies Production Department Hongdong Enterprise Management Company', the 'Shanxi Branch of the Beijing Military Region Logistics Department Coal Transportation and Sale Section', 'The PLA General Political Department Working Committee Shanxi Branch', and a further 63 affiliated organizations. Official seals were seized, in total 164, including 7 steel seals which had been illicitly engraved, as were 2,000 sets of military uniforms and 400 volumes of fake certificates for military officers and identity cards. In a short period of nine months after June 1992, 'Senior Colonel' Guo Rongxi was said to have set up 39 army-run enterprises in 7 provinces and 22 counties and cities, and recruited 400 service personnel, 130 among them being awarded military ranks. From March 1992 to April 1993, 'Colonel' Li Tianyou was alleged to have cheated 15 units and individuals in 11 provinces and cities of money and articles worth 4.8 million yuan through signing counterfeit contracts under which he promised to provide coal and iron to his clients.[33]

Strictly speaking such fraud, if the reports are to be believed, is not an example of corruption within the PLA. The PLA is the excuse and exemplar for a fraudulent exercise, rather than the executor of the fraud itself. However, it is not necessarily the case – even in 1993 – that the Shanxi cases or others like them are self-evidently fraud. It is possible that such cases of fraud appear as ex post facto rationalizations of PLA actions. PLA units may have established subordinate companies in order to market their activities or simply to be entrepreneurial. Such activities may even have crossed the boundaries of the permissable. Policy changes, market failure, or a renewed political campaign may then leave the original PLA progenitors exposed and with no choice but to abandon support for their subordinate companies.

Unfortunately for analysis, both genuine fraud and ex post facto designated fraud are possible interpretations of the available evidence. This ambiguity was particularly acute in an apparent case of 'fraudulent military business' uncovered in Liaoning Province during early 1993. Press reports described a gang of criminals who produced military registration forms, military rank certificates, and merit certificates, and sold them to those who wanted to pass themselves off as demobilized soldiers, which would provide priority in obtaining jobs in the state and collective sectors. Certification as a male soldier cost 10,000 yuan; and as a female soldier, 15,000 yuan. However, those involved in the fraud were most definitely within the PLA: some 300 people were indicted from 43 military units at or above regiment level.[34]

EMBEZZLEMENT

The crime of embezzlement is equally unclear, not least because the regulation of property rights and relationships has dramatically failed to keep pace with economic restructuring.[35] Though appropriations or use of public funds and resources are open to criticism as corrupt behaviour, it is far from certain that all are immoral or corrupt acts let alone illegal, and there is no necessary personal gain. It is estimated that about 100 million yuan a day of state property was appropriated between 1981 and 1993. Though some of this may have become 'privatized' in some sense, and some will almost certainly have been the result of dubious transactions, a goodly proportion will also have been straight transfers to collectives and the collective sector, often of loss-making capital and machinery.[36] At the same time, there clearly is considerable room for corruption particularly on the smaller scale where public servants of all kinds have access to public funds for eating, drinking, and travelling.

Fairly frequent reports of embezzlement within the PLA – the use of military vehicles and other equipment to make money, and even the theft or sale of PLA property – would seem to indicate the ease with which officers and soldiers can move beyond the boundaries of the permissible. These kinds of activities were the subject of severe criticism in a recent speech by the Minister of National Defence, Chi Haotian.[37] Some sources report that as many as 10 per cent of PLA soldiers may be involved in illegal commercial activities, often using military equipment and buildings for such purposes.[38] For example, seventy flats at the military camp of a regiment in Tianjin were reported occupied by the dependents of officers and soldiers in order to engage in trade or work around the camp.[39]

The PLA's *Liberation Army Daily* has admitted that it has recently become a common occurrence for soldiers to steal weapons and other items of military equipment, particularly when they know they can find a ready market.[40] One model division in the Jinan Military Region reportedly found that more than 300 articles of public property disappeared every year.[41] Cars and petrol are particularly marketable commodities. In three months between February and April 1991, the political instructor of a unit of the PLA Navy stole more than four tons of petrol from a military oil terminal and sold it for 4,400 yuan.[42] On 28 May 1993 a PLA driver stole a car from his local headquarters and sold it for some 130,000 yuan. During the period from March to August 1993 six PLA drivers supplied military vehicles that were used for robberies on more than twenty occasions.[43]

Like their civilian counterparts, some PLA officers would appear to have abused their access to public funds for eating, drinking and seeking pleasure, especially when they are on tours of inspection or receiving personal guests. For example, a motor vehicle materials storekeeper of one military academy is reported to have embezzled 11,523 yuan between October 1991 and September 1992 by filling in false receipts for reimbursement.[44] This practice has been so widespread that it has been the target of its own specific propaganda campaign. A regimental Deputy Commander was praised by *Liberation Army Daily* for refusing to engage in such practices. He had spent 100 yuan on entertaining a personal dinner guest at regimental headquarters. As usual in such circumstances, the kitchen staff prepared an exaggerated receipt so that reimbursement could be claimed, but he tore it up.[45] On 6 September 1993 the CCP committee of a regiment in the Beijing Military Region decided not to serve guests with 'famous-name liquors, famous-name cigarettes, and rare delicacies from the mountains and the sea'. This decision was also hailed as an exemplary deed by the PLA newspaper.[46]

SMUGGLING

Smuggling is both a more specific crime and one in which the PLA plays a central role. Even though information on smuggling is largely only derived from the PRC media it would seem that a substantial number of service personnel and military units are engaged in this activity. In August 1993 the PRC National Work Conference on Cracking-down on Smuggling pointed out that smuggling activities in the country were growing rapidly in size, extent and seriousness. Though its criticisms of the perpetrators was considerably more muted than its descriptions of specific acts, the conference nonetheless fired a warning at local authorities and units of the PLA engaged in or supporting smuggling activities. Local authorities were accused of using administrative means to shield smuggling activities to a serious degree, and almost all the transport for smuggled goods was said to be military stock: including naval cargo ships, military landing boats, special trains for military use, and military vehicles of all kinds.[47]

Maritime smuggling has spread from the southeast coastal areas northwards to all coastal areas. Smuggling along land borders was also now considered to have increased dangerously, and was expanding in various trade forms. The number of cases of smuggled cars had increased sharply, with at least 26,522 cars from the Republic of Korea and 27,000 cars from Hong Kong. It was reported that 1.14 million colour television sets and 510,000 video recorders had been smuggled into China from Japan in 1992. The number of large-scale and serious cases, especially those in which enterprises and institutions were incriminated, had also increased dramatically with 324 cases involving over one million yuan each among the 376 major cases cracked down on in the first half of 1993.[48]

PLA involvement in smuggling is not only a function of its national defence role and hence its physical location on the borders, but also its ready access to considerable and varied amounts of transport, and its key role in public security. The PLA at the provincial and sub-provincial levels is largely engaged in public security as opposed to military activities, and has a leading role in local public security activities. Apart from any other consideration this provides a certain amount of protection to illegal operations such as smuggling, not least since PRC Customs anti-smuggling teams cannot investigate a military unit without special approval from Beijing. The PLA's senior position among the public security organizations – the 'eldest brother' – not only makes investigators hesitate, it also directs those interested in smuggling towards PLA and service personnel for assistance.[49]

However, PLA units and soldiers have often been centrally involved in smuggling activities as organizers and initiators. An early and much publicized case was related to the famous and quasi-legal importing of foreign cars by Hainan Island in the summer of 1985 when local government used its developmental status to buy foreign cars that were then sold on to within-PRC purchasers without use on Hainan. At that time some gunboats of the PLA Navy South China Sea Fleet were involved in the smuggling of cars out of Hainan in cooperation with the Hainan local government. A more recent, but almost equally celebrated, example was the capture off Jiazi in Guangdong during the summer and autumn of 1992 of 38 'ex-service' gunboats, torpedo boats, escort vessels, and submarine chasers, all involved in smuggling.[50]

Though both the Hainan and Jiazi cases received considerable coverage in the media outside China, there are many further examples of PLA involvement in smuggling. In August 1988 the Logistics Department of an airforce unit bought fifteen smuggled Japanese cars at Dongguan City in Guangdong, broke them deliberately into pieces – both to avoid detection and to conform to import regulations – and transported them to Henan Province.[51] In September 1988 another unit's Logistics Department bought three smuggled Japanese cars from the Nanning Military District in Guangxi. The cars were also broken into pieces in order to be carried legally to northern China, but on the way were seized by Customs officials in Hunan.[52]

By all accounts both the scale and potential for catastrophic consequences of the PLA's smuggling activities appear to have grown considerably during the 1990s. Starting in June 1993, the Logistics Department of the PLA Navy North China Sea Fleet appears to have launched an operation designed to smuggle some 2,200 cars and vans into China from the Republic of Korea and Russia, by stages and in batches. Unfortunately for the smugglers, the fourth group was captured on 9 August at sea by Customs officials who were acting on special instructions approved by the PRC State Council. In mid-August Minister of National Defence, Chi Haotian, went to Tianjin to handle the case personally.[53] In May 1990 it was reported that a PLA Navy patrol boat which was escorting a boat engaged in smuggling opened fire on an anti-smuggling patrol boat of the local public security office off the coast of Shantou. Fire was exchanged, resulting in more than twenty deaths.[54] In November 1990 the Beijing Customs Administration uncovered an operation to smuggle weapons and drugs. Among the eight who faced criminal proceedings when arrested, five were children of senior army officers, one of whom was a deputy chief of the general staff of the PLA.[55] Though there was no direct PLA involvement cited in this case, there were in the end more than 700 cases of arrests for drug smuggling at and above sectional level within the PLA during 1990.[56]

CONSEQUENCES

The very nature of corruption makes it hard to assess its scale or extent in general, and this problem of observation is magnified by the security shield surrounding the PLA. There clearly is corruption of various sorts within and involving the PLA – or at least its constituent units and personnel – and the arguments presented here about the causes, contexts, and consequences of corruption are in no way intended to deny its existence or provide apologia. However, there may well be a difference between corrupt actions and the CCP's recognition of corruption, and the causes of either may be less the moral malaise identified by the PLA and CCP than structurally determined.

The arguments that corruption may in certain circumstances not be dysfunctional, and particularly that which suggests that corruption may redistribute public goods as society becomes more open, should also not be confused with an argument for greater equality. One interesting feature of corruption within the PLA is that it appears uneven: not everyone within the PLA benefits, hence the complaints even from within its own ranks. *Liberation Army Daily* has complained loudly that it has become a widespread phenomenon to 'materialize' personal feelings in the hierarchical relationships between officers and soldiers.[57]

Soldiers mock the corrupt officers as 'having irregular features: weak eyesight incapable of seeing problems, deaf ears incapable of hearing complaints, lazy legs that never walk on the exercise ground, long hands that grab at everything, and a greedy mouth that eats a lot'.[58]

The CCP and the leadership of the PLA are concerned that corruption may in some sense weaken the PLA as an institution and the PRC's national defence policy. Even if corruption is structurally rather than morally determined that concern is obviously justified. Some engagement by PLA units and personnel in some – though by no means all – of the activities that the CCP may from time to time criticize as economic corruption may reinforce morale and well-being in an environment of rapid socio-economic change, and paradoxically even strengthen the PLA as an institution under such circumstances. However, if such activities came to be the major preoccupation for most PLA units and personnel then its military capacities would almost certainly suffer. Given the past emphasis on ideological purity – both for the PLA and China as a whole, historically – the CCP's appeal to morality is thus probably the most effective strategy for maintaining the boundaries of the permissible.

NOTES

1 For details see, *Internal Documentation of Central Commission for the Inspection of Discipline*, 10 March 1993, translated in *Inside China Mainland*, September 1993, p. 18.
2 Zhang Siqing, 'Jianjie chenzhi tanwu huilu deng jingji fanzui' ('Resolutely punish corruption and other economic crimes') in *Renmin Ribao* (People's Daily), 1 November 1993, and 'Ideals and realities of politics' in *Liaowang* (Outlook Weekly), no. 26, 28 June 1993, p. 9; translated in *Inside China Mainland*, September 1993, p. 14.
3 *Renmin Ribao* (People's Daily), 23 August 1993, p. 1.
4 Liu Huaqing and Zhang Zheng, 'Fayang youliang chuantong shi xin xingsi xia wo jun jianshe de zhongda zhanluexing wenti' ('It is a question of vital strategy to uphold the good tradition of our army in the new situation') in *Jiefangjun bao* (Liberation Army Daily), 26 July 1993.
5 See, for example: Keith Forster, 'The 1982 Campaign against Economic Crime in China' in *Australian Journal of Chinese Affairs*, no. 14, p. 1.
6 James C. Scott, *Comparative Political Corruption Affairs*, Englewood Cliffs, New Jersey, Prentice Hall, 1972, p. 4.
7 See, for example: Jonathan Unger, 'Rich Man, Poor Man: Socio-economic trends in the making of new classes in the Chinese countryside' in David S. G. Goodman and B. Hooper, *China's Quiet Revolution: New Interactions between State and Society*, Melbourne, Longman Cheshire, 1994.
8 An excellent argument of this kind may be found in Stephen K. Ma, 'Reform

Corruption: A Discussion on China's Current Development' in *Pacific Affairs*, vol. 62, no. 1, Spring 1989, p. 40.

9 See, for example: Liu Huaqing and Zhang Zheng, 'Fayang youliang chuantong shi xin xingsi xia wo jun jianshe de zhongda zhanluexing wenti' ('It is a question of vital strategy to uphold the good tradition of our army in the new situation') in *Jiefangjun bao* (Liberation Army Daily), 26 July 1993.

10 Chen Yan and Liu Siyang, 'Ningju dangxin minxin, zhongjiwei er ci quanhui ce ji' ('The party and the people come together: a report on the second meeting of the CCP Central Commission for Inspection of Discipline') in *Renmin Ribao* (People's Daily), 24 August 1993; and 'Over half a million party members disciplined in past three years' in BBC, *Summary of World Broadcasts* (SWB) FE/1772, B2/7, 20 August 1993.

11 '*Cheng Ming* gives figures for corruption among PRC officials' in *SWB* FE/1658, B2/2, 8 April 1993.

12 'Official Continues Fighting Corruption' in *China Daily*, 22 December 1993.

13 'Procuratorate handles over 27,000 cases of corruption in first half 1993' in *SWB* FE/1795, G/4, 16 September 1993.

14 Zhang Siqing, 'Jianjie chenzhi tanwu huilu deng jingji fanzui' ('Resolutely punish corruption, bribery and other economic crimes') in *Renmin Ribao* (People's Daily), 1 November 1993; and Xie Gong, 'Jiusan Zhongguo fanfu changlian fengyun lu' ('The anti-corruption storm in China in 1993') in *Minzhu Yu Fazhi* (Democracy and the legal system), no. 10, 1993.

15 See, for example: Liu Huaqing and Zhang Zheng, 'Fayang youliang chuantong shi xin xingsi xia wo jun jianshe de zhongda zhanluexing wenti' ('It is a question of vital strategy to uphold the good tradition of our army in the new situation') in *Jiefangjun bao* (Liberation Army Daily), 26 July 1993.

16 On the recent development of military businesses in China, see Arthur S. Ding, 'The Nature and Impact of the PLA Business Activities' in *Issues and Studies*, vol. 29, no. 8, August 1993; and Mel Gurtov, 'Swords into Market Shares: China's Conversion of Military Industry to Civilian Production' in *The China Quarterly*, no. 134, June 1993.

17 See, for example: Tai Ming Cheung, 'Profits over Professionalism: The People's Liberation Army's Economic Activities and its Impact on Military Activity' – paper delivered to the Conference on Security Dimensions of Chinese Regionalism, International Institute for Strategic Studies and Chinese Council of Advanced Policy Studies, Hong Kong, June 1993.

18 'Junshi fayuan shenli jingji jiufen anjian de ruogan wenti' ('Several questions on handling cases of economic disputes by military tribunals') in *Jiefangjun bao* (Liberation Army Daily), 4 January 1993; and Geng Renwen, 'Junnei jingji shenpan yin zhunxun weihu junshi liyi yuanzhe' ('The trial of economic cases in the army should follow the principle of protecting military interests') in *Jiefangjun bao* (Liberation Army Daily), 31 May 1993.

19 Gen Renwen, 'Junnei jingji shenpan yin zhunxun weihu junshi liyi yuanzhe' ('The trial of economic cases in the army should follow the principle of protecting military interests') in *Jiefangjun bao* (Liberation Army Daily), 31 May 1993; and Arthur S. Ding, 'The Nature and Impact of the PLA's

Business Activities' in *Issues and Studies*, vol. 29, no. 8, August 1993, p. 95.

20 Mel Gurtov, 'Swords into Market Shares: China's Conversion of Military Industry to Civilian Production' in *The China Quarterly*, no. 134, June 1993, p. 240.

21 Huang Bensheng, 'Zenyang kan fancha?' ('How should we treat the disparities?') in *Jiefangjun bao* (Liberation Army Daily), 20 February 1993.

22 Yao Jianping, 'Gaige ye honghuo, tounao ye qingxing' ('The more fervent the reform becomes, the clearer head we should keep') in *Jiefangjun bao* (Liberation Army Daily), 9 May 1993.

23 Yang Baibing, 'Fan fushi shi jijian gongzuo de zhongyao renwu' ('Anti-corruption is an important task of discipline inspection work') in *Renmin Ribao* (People's Daily), 24 March 1988.

24 Xu Mingye, 'Wenti zao zhua gongzuo zuoshi: mou shi yifang ganbu tanwu huilu fanzui de qishi' ('Handle the problem early and solve it carefully: inspiration drawing from the experience of preventing cadres from corruption in a certain division') in *Jiefangjun bao* (Liberation Army Daily), 19 October 1992.

25 Jie Yanzhen, 'Bu wei qiancai shi bense: ji Nanjing junqu houqin bumen yixie lingdao ganbu' ('No sacrifice of inherent quality for money: a report on some cadres of the logistics departments in the Nanjing Military Region') in *Renmin Ribao* (People's Daily), 3 December 1988.

26 Guo Aishu and others, 'Mindui chuanli de kaoyan' ('Facing the test of power') in *Jiefangjun bao* (Liberation Army Daily), 26 September 1993; Huang Shangsong and Liao Zhuangqiong, 'Chuan bu lanyong bianyu bu zhan: jiefangjun moubu dangwei lianjie zilu' ('Not abusing power for personal gain: the party committee of a certain unit of the PLA disciplines itself') in *Renmin Ribao* (People's Daily), 31 August 1993.

27 Wu Weijun and Yao Bolin, 'Guanghou jijian bumen zhichi junban xuye zouxiang shichang' ('Discipline Inspection Branches of the Logistics Department of the Guangzhou Military Region support military enterprises to enter the market') in *Jiefangjun bao* (Liberation Army Daily), 28 January 1993.

28 Geng Shunxi and Zhou Li, 'Gongdao banshi nanshi bu nan' ('There will be no difficult problems if there is fairness') in *Jiefangjun bao* (Liberation Army Daily), 3 February 1993; Xu Xianrong and Zheng Zongxun, 'Quanzhoushi zhidui dangwei lianzheng jianshe you mai xin taijie' ('A new step taken by Quanzhou City Forces Party Committee in building a clean administration') in *Jiefangjun bao* (Liberation Army Daily), 4 October 1992.

29 Dong Wu, 'Tuibian zhong de zhonggong jundui' ('The PLA in decay') in *Zheng Ming*, no. 124, February 1988.

30 Li Xusheng, Li Quanmou and Tao Ke, 'Xin shiqi gongfu de meili' ('Charm of the public servants in the new era') in *Jiefangjun bao* (Liberation Army Daily), 14 December 1992.

31 Zhang Shunlin, 'Bunengchu jie gongzhang' ('Don't lend the official seals') in *Jiefangjun bao* (Liberation Army Daily), 17 May 1993.

32 'Jiangxi sheng junqu dangwei weizheng qinglian fengqi zheng' ('The party committee of the Jiangxi Province Military District is honest and upright in politics') in *Renmin Ribao* (People's Daily), 7 November 1988.

33 Wen He, Zhengzhe and Linxiang, 'Jie bu yunxu dingwu shengjie de junhui: Shanxi teda jiamao junren tuanhuo xianxing ji' ('Never stain the holy and pure army emblem: discovery of exceptionally large false service personnel gangs in Shanxi') in *Jiefangjun bao* (Liberation Army Daily), 18 September 1993; 'Shanxi rounds up gangs running fake military enterprise' in *SWB* FE/1800, G/6, 22 September 1993.
34 'False Soldiers Business Discovered in Liaoning' in *SWB* FE/1819, G/4, 14 October 1993.
35 'The necessity of straightening out property rights' in *Jingji Guanli* (Economic Management), no. 9, 1993, p. 36; David S. G. Goodman, 'The Political Economy of Change' in D. S. G. Goodman and B. Hooper (eds), *China's Quiet Revolution: New Interactions between State and Society*, Longman Cheshire, Melbourne, 1994, pp. 19ff.
36 Zhan Guoshu, 'Shui lai baohu guoyou zichan' ('Who will protect state property?') in *Jingji Ribao* (Economic Daily), 1 May 1994.
37 Li Zhu, 'Chi Haotian qizhi chuli jundui zhoushi an' ('Chi Haotian personally handles a smuggling case in the army') in *Zheng Ming*, no. 191, September 1993.
38 'Iniquitous Inequities', *Junshi jingji yanjiu* (Military Economic Studies), no. 6, 1990, p. 76, translated in *Inside China Monthly*, vol. 12, November 1990, p. 26.
39 Wang Guansheng and Shi Dali, 'Yixie guan bing jiashu jiliu yinqu dagong jingshang: mou tuan linshi jiashu yuan baoman' ('Some dependants of the officers and soldiers stay at the camp to engage in trade and other work: the armymen's families compound of a certain regiment is over-crowded') in *Jiefangjun bao* (Liberation Army Daily), 14 April 1993.
40 Gao Laifu and Liu Xianqiang, 'Jiaqiang dui junren weifa fanzui shehui youyin de fangfan' ('Strengthen guard against temptation for soldiers to commit crimes') in *Jiefangjun bao* (Liberation Army Daily), 17 May 1993.
41 Xu Mingyue, 'Wenti zao zhua gongzuo zuoshi: mou shi yifang ganbu tanwu huilu fanzui de qishi' ('Handle the problem early and solve it carefully: inspiration drawing from the experience of preventing cadres from corruption in a certain division') in *Jiefangjun bao* (Liberation Army Daily), 19 October 1992.
42 Zheng Jingjie, 'Zhiquan shengliao ta moushi de jieti' ('Power becomes a tool for him to seek personal gain') in *Jiefangjun bao* (Liberation Army Daily), 14 December 1993.
43 'Jidai jiaozheng de fangxiang pan' ('The steering wheels need urgent readjustment') in *Jiefangjun bao* (Liberation Army Daily), 4 December 1993.
44 Zhang Lin and Ye Sanfang, 'Fapiao shangde jia wenzhang' ('The conspiracy of receipts') in *Jiefangjun bao* (Liberation Army Daily), 4 January 1993.
45 Li Xucheng, Li Quanmou and Tao Ke, 'Xinshiqi gongfu de meili' ('The charm of public servants in the new era') in *Jiefangjun bao* (Liberation Army Daily), 14 December 1992.
46 'Moutuan dangwei zhua lianzheng xian xiang zhiji kai sandao' ('A certain regiment party committee makes an example of itself in incorrupt management') in *Jiefangjun bao* (Liberation Army Daily), 26 September 1993.
47 Liu Tian, 'Zhonggong jingtan dalu sheng zousi wangguo' ('The CCP cries

out that Mainland China has become a smuggling kingdom') in *Zheng Ming*, no. 191, September 1993.

48 'Article surveys rise in smuggling across land, sea borders' in *SWB* FE/ 1824, G/19, 20 October 1993.

49 Peng Baowu, 'Haijun moubu jianchi yuanzhe weihu guojia liyi, ju daoye yi junying zhiwai' ('A certain unit of the PLA Navy sticks to the principle of protecting the state interests and keeps speculators out of the barracks') in *Renmin Ribao* (People's Daily), 19 November 1988; Chen Lianjun and Xiangyang, 'Lingdao ju fushi budui fengqi zheng: zhu kouan mou shi guanbing jilu yanming shou biaoyang' ('Leading cadres resist corruption and keep the general mood of troops healthy: a certain division stationed at the ports is praised for maintaining good discipline') in *Jiefangjun bao* (Liberation Army Daily), 13 July 1993. Xiao Xiaozhi and Ye Jing, 'Huangjin tongdao zhu liba' ('Build the fence to block the "Golden Road"') in *Jiefangjun bao* (Liberation Army Daily), 13 September 1993.

50 For further details see: Yan Changjiang, *Guangdong Da Liebian* (The disintegration of Guangdong), Jinan University Press, 1993, p. 242.

51 'Junwei zhongshi yulun piping chashu jingji anjian jie bu huduan' ('The Central Military Commission pays attention to criticism by mass media and deals sternly with the cases of economic crime) in *Renmin Ribao* (People's Daily), 28 February 1989.

52 Du Royuan, 'Dongyong jinche junche youche chuangguan zousi fanshi daomai jinkou qiche' ('Use police cars, military vehicles, and postal cars to halt the smuggling of imported cars') in *Renmin Ribao* (People's Daily), 5 October 1988.

53 Li Zhu, 'Chi Haotian qizhi chuli jundui zhoushi an' ('Chi Haotian personally handles a smuggling case in the army') *Zheng Ming*, no. 191, September 1993.

54 He Zhengming, 'Bianfang shaojiang tequan jingshang' ('A major-general of the Border Force engages in business in a privileged position') in *Zheng Ming*, no. 152, June 1990, p. 24.

55 Qiu Zhixing, 'Fuzongzhang zi zi zousi wuqi duping' ('Son of Deputy Chief of the General Staff involved in smuggling weapons and drugs') in *Zheng Ming*, no. 160, February 1991, p. 16.

56 Chen Zijian, 'Qibai guanyuan fanshi beibu' ('Seven hundred officials arrested for drug trafficking') in *Cheng Ming*, no. 161, March 1991, p. 92.

57 Jing Shengzhi, 'Zhengque yindao shichang jingji tiaojian xia de junyin renji guanxi' ('Give correct guidance to interpersonal relationships in barracks under conditions of a market economy') in *Jiefangjun bao* (Liberation Army Daily), 14 July 1993.

58 Lu Jianrong and Zhou Jintao, 'Zheng wuguan de bingxin' ('Correcting the five organs and winning good feelings from soldiers') in *Jiefangjun bao* (Liberation Army Daily), 8 July 1993.

4 'PLA incorporated'

Estimating China's military expenditure

Paul H. B. Godwin

Determining what Beijing spends on defence is an impossible task. This chapter does not pretend to do so. Rather, its purpose is to explore the problems involved in estimating Beijing's military expenditures. That said, the simple matter of language used to describe these expenditures can be confusing. For the purposes of this chapter, the term defence budget refers to the amount specifically allocated to national defence in the Minister of Finance's annual state budget report. Wherever it is known, tables in this chapter refer to the funds actually expended on defence as reported in the annual budget statement. Usually, the funds disbursed are larger by a small fraction than the amount allocated. Military expenditures will refer to funds and resources used for defence beyond those specified in the budget as spending for national defence.

Widespread concern over Chinese military expenditures, especially in Asia[1] and the United States,[2] stems from the rapid increase in defence budget allocations over the past five years. In essence, it appears that as China's economy has grown significantly richer the military has begun to reap the results of the PRC's economic dynamism. On its own, increasing Chinese defence budgets is not necessarily a cause for concern. The apprehensions that have emerged, however, are a direct result of three factors. First, the lack of transparency evident in Beijing's defence budget reports. The defence budget consists of a single-line entry with no breakdown into operations and maintenance, personnel costs, research and development, etc. In terms of personnel, the Chinese armed forces are the world's largest. When personnel, weapons and equipment are combined, the Chinese People's Liberation Army (PLA – as the combined Services and branches of the Chinese armed forces are designated) ranks behind only the United States and Russia in overall force size. With personnel exceeding 3 million manning 24 Group Armies, over 5,500 fixed-wing combat aircraft, around 60 naval combatants exceeding 1,000 tons displacement, strategic and theatre nuclear forces deploying some 14 inter-continental ballistic missiles (ICBMs), 60 intermediate range ballistic missiles (IRBMs), 1 nuclear submarine with ballistic missile launching capability (SSBN) with 12 missile tubes, and an unknown number of air-delivered weapons,[3] a defence budget of around US$6.0 billion[4] cannot represent the cost of sustaining such a military capability. When China's ongoing modernization programmes for both its conventional and nuclear forces are brought into focus, the defence budget can be viewed as more than merely lacking in transparency – some will see its primary purpose as deception.

The second factor is the aggressive attitude Beijing takes toward its sovereignty claims over the Paracel (Xisha) and Spratly (Nansha) islands in the South China Sea. China's claims are contested in total or in part by Vietnam, Malaysia, the Philippines, and Brunei. Taiwan, of course, presses precisely the same claim as Beijing. The uncompromising position taken by Beijing is seen as a harbinger of future Chinese policy as its military begins to reap increasing financial resources from the growth of China's booming economy.

The third factor is the developing relationship between the Chinese and Russian defence establishments. China's policy of limiting its defence expenditures has been frequently and loudly stated. Nonetheless, Beijing's purchase of 26 SU-27s (the Russian equivalent of the US F-15 Eagle), sophisticated air defence missiles, and presumed expanding military technology link with Moscow are viewed by many in the region as demonstrating Beijing's intent rapidly to develop its military might as China's fast economic growth permits larger allocations to defence spending. The poverty of Russian military defence industries demands that they seek an expanding market, and China is a willing partner with both the convertible currency and barter goods Russia needs.

Yet, China's Minister of Foreign Affairs is joined by the senior members of Beijing's national security élite in stating that the PRC's military expenditures are what the state budget says they are. What is more, directors of the PLA's General Logistics Department (GLD) have argued for the past five years that China's military expenditures are as they are reported in the state budget, and that these funds are so limited that even with annual increases reaching into double-digit percentages the military establishment can barely sustain itself. These observations are supported by articles in Chinese journals contending that limitations on defence spending have not only impoverished the armed forces – they have severely limited the upgrading of the PLA's weapons and equipment. Essays describing the poverty of the armed forces are clearly seeking a larger allocation of resources to China's armed forces in order to implement the defence modernization programmes promised by Deng Xiaoping at the Third Plenum of the Eleventh Central Committee in December 1978. Modernization of national defence is one of the 'Four Modernizations' that form the core of Deng's strategy to build China into a powerful state in the twenty-first century, albeit the fourth in terms of resource investment priorities.

The gap between what many observers perceive as actual Chinese military expenditures and what is reported in the state budget is the focus of this chapter. Quite clearly, statements by Chinese officials and analysts that the PRC's military expenditures are minimal and among the lowest in the world are not accepted by the world at large. The dilemma this chapter faces is how to bring these contradictory views into balance, for I do not believe that Chinese officials and analysts are deliberately masking a massive defence build-up through which the PRC intends to dominate Asia. Thus, before entering into an analysis of Beijing's military expenditures it is essential to provide an overview of the defence modernization programme underway since December 1978, and the problems that emerged in the late 1980s and continue until this day.

CHINA'S DEFENCE MODERNIZATION

It would not be too much of an exaggeration to define the Chinese armed forces as the world's largest junkyard army.[5] Most of the weapons and equipment they deploy are based on Soviet military technologies of the 1950s. Only in recent years, especially since the early 1980s, has Beijing embarked on a systematic programme of modernization to turn a vast, sprawling, technologically retarded military into something like a modern defence force. Mao Zedong, for his own ideological and political reasons, brought the first attempt to modernize China's armed forces with the assistance and tutelage of the Soviet Union to a close in 1959, barely four years after the programme was initiated shortly after the Korean war. The Sino–Soviet split and Beijing's foreign policy left China isolated from the mainstream of military technology and military expertise until the early 1980s. Forced to rely on the PRC's limited scientific and technological capabilities, Mao Zedong focused what few resources China had on the nuclear weapons and ballistic missile programmes. The conventional forces had to rely on reverse-engineering Soviet weapons and equipment for domestic production. Series production of these weapons did not begin until the mid-1960s. Equally if not more damaging to the combat capabilities of the armed forces, Chinese military personnel became deeply involved in Mao Zedong's domestic political campaigns leading to and including the Great Proletarian Cultural Revolution. As a consequence, at a 1975 expanded meeting of the Central Military Commission of the Chinese Communist Party (CMC), Deng would declare that the Chinese armed forces were bloated, arrogant, ill-equipped, and too poorly trained to conduct modern warfare.[6] The performance of the PLA in its 1979 border clash with Vietnam appears to have validated Deng's judgment.

Deficiencies within the Chinese armed forces were found not only in their obsolescent equipment, but also in their training. The Chinese armed forces did not train for combined arms warfare, and their officer corps did not progress through a systematic programme of professional military education (PME) designed to prepare them for higher levels of command, planning and staff duties. Indeed, there was no 'officer corps' as this term is commonly understood. Officer ranks introduced for the first time during the Soviet-assisted modernization programme of the 1950s were eliminated for ideological reasons in 1965, and not restored until 1988. 'Cadres', as the PLA's officer equivalents were known, were recruited primarily from the ranks of conscripts. Upon completion of their conscripted service, usually three or four years, selected young men and women were invited to become 'volunteers' and to enter into the cadre ranks. Typically, promotions occurred within units, meaning that cadres stayed within the same military district or region and within the same regiment, division, or corps for their entire military career. What is more, there was no mandatory retirement age, therefore cadres could remain in the same command or staff position until they chose to depart the armed forces. Thus, military units were often led and staffed by cadres with very little experience outside their own military component. One could conclude from the criticism directed at the PLA by its own reformers that the Chinese armed forces entered the 1980s poorly equipped, poorly trained, and poorly led.

When Deng Xiaoping included the 'modernization of national defence' as one of the Four Modernizations, his focus was therefore far broader than the armed forces themselves. Deng was seeking to modernize China's entire defence establishment: the armed forces and their training and professional development, communications infrastructure, the defence industrial base, and defence research, development, testing and evaluation (RDT&E) capabilities. Given the decrepit state of each of the defence establishment's components, this was not a task that could be completed within a decade or even two decades. For Deng Xiaoping, modernization of national defence, like the other three elements of his Four Modernizations programme, would take thirty or more years to complete.

Modernizing China's armed forces quickly assumed a clearly observable pattern that followed short-term and long-term policy objectives. For the short term, the armaments deployed by the armed forces would be improved by upgrading the weapons and target acquisition systems used on existing platforms. To the extent possible, the platforms themselves would also be improved. In both cases, foreign technology would be used but domestic production of imported systems was preferred over the direct importation of end-use items. Western weapons and equipment were very expensive and purchasing these items would require expending scarce convertible currencies. Furthermore, domestic production would improve the capabilities of the defence industries to place advanced technology systems into series production while allowing the re-equipment of far more military units than would occur if end-use items had to be purchased for each weapon platform. This process of selected purchase of end-use items and technology was accompanied by the slimming down of the armed forces through a one million-man reduction-in-force (RIF) instituted in 1985, the reorganization of the armed forces, a radical change in the armed forces' training programme to achieve combined arms warfare capabilities, and the reconstruction of the officer corps.

Long-term modernization objectives were wrapped into the short term, but were distinct. Weapons and equipment modernization was to depend on the development of a defence industrial base capable of series production of advanced technology weapon systems and military equipment. Licensed (and unlicensed) production of weapons and equipment purchased from the West would permit the defence industries to develop such a capability. Similarly, technology purchased from the West and integrated into weapons and equipment production would provide the defence industries with valuable experience. Improving the quality of China's science and technology (S&T) personnel through training at the universities of advanced industrial states would provide an infusion of young, well-trained engineers and scientists into both civilian and military research institutes.

The recruitment of new officers primarily from among college graduates, or selected from conscripts who were then sent to military colleges before commissioning into the armed forces, would provide the armed forces with well-educated officers capable of being trained in the use and maintenance of high technology weapons and equipment as more modern systems were introduced over time. These same officers would be developed through a systematic process of professional military education (PME), with each officer required to complete a specific level of PME prior to promotion and reassignment.

An integral part of modernizing the defence establishment was the slimming down and reorganization of the defence industrial base and defence science and technology (S&T) centres. Beginning in 1982, the defence industries were removed from military control and placed under civilian directors. Furthermore, they were instructed to reduce their dependence on military orders by diversifying into products for the civil sector of the economy. For civilian products, the defence industries were to depend upon market forces for both the materials and the products they were to sell. In this sense, 'defence conversion' in China meant not only converting from military to civil products, it also meant conversion from the command economy to the market economy. Only military orders were to come under the direction of central planning along with the supply of materials to be used in military products. Similarly, the defence S&T community was integrated into the civil S&T structure. Research institutes were no longer required to isolate themselves from each other, and every effort was made within the constraints required by the classification of their projects to ensure the exchange of results and data. Similarly, where suitable, defence technology was to be used in the production of products for the domestic market and for export.

Deng's strategy to transform China's lumbering, obsolescent defence establishment was premised on a world where the Soviet Union and the United States continued to preoccupy one another. In such a world, China's security analysts predicted that Moscow and Washington would gradually lose their predominance to a multipolar pattern of global politics. These analysts concluded the military threat to China was minimal and that the dynamics of the international system gave Beijing the opportunity to develop the economic, scientific, and technological capabilities essential for China's emergence as a major power in the twenty-first century. In the early summer of 1985, the CMC modified China's national military strategy to reflect the minimal threat of a major war. The CMC declared that it was no longer necessary to prepare for a major and possibly nuclear war. China's new national military strategy was based on preparing for possible limited wars or unanticipated military conflicts on China's periphery. The PLA was charged with developing a military strategy and concepts of operations to respond to such contingencies.

As the armed forces adjusted to the new limited war strategy, the PLA stressed the need for quick, lethal responses to military crises. Modern limited war operations, Chinese military strategists noted, use high-technology weapons and equipment in fast-paced combat. Quick-reaction forces were essential, and these forces had to contain the capability to defeat adversaries early in the conflict. There would be no time to mobilize the entire nation for war. The armed forces would be required to move with great speed and lethality if they were to achieve their political and military objectives. Such a requirement placed great emphasis on mobility and military technology. The PLA was deficient in both areas. *Quantou*, or 'fist units', were developed to act as quick-reaction units. They were provided with the most modern equipment available to the PLA. The task of these units, in addition to being the forces in readiness for military emergencies, was to develop tactics and the operational concepts for the use of the PLA's latest equipment and to gain experience in field maintenance. Such experience would be passed on to other units as the modern equipment became more available. Quick-reaction units were developed within each of China's seven military regions.

Even as this new strategy and the supporting concepts of operations were being digested, the tragedy at Tiananmen Square occurred. Beijing's brutal response to the demonstrators and the consequent suppression of dissidents drastically changed the West's image of China. No longer viewed as a model for transforming Marxist–Leninist societies, Beijing's aging political élite were seen as a throwback to the worst days of Stalin and Mao. The collapse of the Berlin wall and the ultimate disintegration of the USSR followed by Russia's emergence as a proto-democratic system highlighted China's political isolation as the world's last remaining Communist state of any significance. Within Beijing, the global events following on the heels of the Tiananmen crisis only heightened the sense of isolation pervading China's political élite within Zhongnanhai. When the United States formed the coalition that ejected Iraq from Kuwait, security analysts in China perceived the multipolar world they once saw emerging become transformed into a unipolar system led by the West and the United States.

Even before the Gulf War, Chinese military exercises evaluating the armed forces' quick reaction capabilities were accompanied by the PLA's demand for more rapid updating of its weapons and equipment. Since the late 1970s, the requirements of modern warfare have been cited by Chinese analysts as demonstrating the need for modern weaponry. By the late 1980s, a new argument was being introduced: the defence budget lagged so far behind the rate of inflation accompanying China's rapid economic growth that the planned modernization of weapons and equipment had not occurred. This argument was presented in Chinese military journals and it became quite evident that considerable effort was being placed on economic arguments to obtain increases in the defence budget. Thus, the dramatic demonstration of high-technology weaponry and equipment during the Gulf War provided yet another arrow in the PLA's quiver of demands. As deputies to the Fourth Session of the Seventh National People's Congress (NPC) in March, 1991, General Liu Huaqing, Senior Vice-chairman of the CMC, and Defence Minister General Chi Haotian both contended that as China's economy improved additional funding had to be allocated to national defence. Both stated that the armed forces' weapons and equipment had to be improved and that China must grant priority to defence S&T.[7]

By 1990, a consensus had emerged within the Chinese armed forces that Deng's strategy for defence modernization contained a flaw. In the late 1970s, the leadership of China's armed forces had accepted, albeit reluctantly, placing national defence modernization last in Deng's overall priorities. After a decade of reform and reorganization within the armed forces, however, the military leadership was faced with a new dilemma. Military technology was a moving train. Deng's interim strategy of improving the PLA's weapons and equipment by upgrading systems based on technologies from the 1950s and 1960s did not permit the armed forces to get on the technology train. The Gulf War seemed to demonstrate as no other war had the profound implications of high-technology warfare when conducted by well-trained armed forces. The Gulf War demonstrated to China's military leadership that the position they had taken in earlier years was correct.

Deficiencies in the armed forces' combat readiness and effectiveness were not the only problems the PLA had to face. Lean defence budgets that failed to keep pace with inflation had led to a degradation of soldiers' living conditions while simultaneously providing officers with incentives to leave the army and seek their fortunes in China's dynamic market economy. This latter inducement was simply heightened by the experience many staff officers had gained in operating PLA enterprises such as hotels, trading corporations, and arms sales. *Xia hai*, or swimming in the ocean of the market economy, became the catchphrase for officers departing the armed forces in search of lucrative business opportunities.

By the late 1980s, and certainly by 1991, Deng's strategy for transforming the Chinese defence establishment was facing what has to be recognized as a crisis; a crisis that could not be resolved by ideological demands. The vaunted PLA that once stood as China's 'Great Wall of Steel' was suffering from severe corrosion. Increasing defence budgets were therefore a response to a variety of ills, only one of which was the obsolescence of the PLA's weapons and equipment that became so magnified by the Gulf War.

THE DEFENCE BUDGET[8]

From 1978 to 1986, three-year budget plans were the norm,[9] but the defence budget is now based on five-year plans with annual budget submissions. The General Logistics Department (GLD) of the PLA is responsible for building the annual defence budget, with final authority for the proposed budget in the hands of the party Central Military Commission (CMC) and the State Council. The GLD compiles the budget in consultation with the General Staff Department (GSD), the General Political Department (GPD), the individual service headquarters and the Military Region headquarters. The process appears to proceed as follows:

- Early in the year, military units down to the division level assess their fiscal requirements.
- Staff departments and military region logistics departments submit their budgetary requests to the GLD in the autumn. These requests are reviewed by the GLD, which makes changes and adjustments as it sees fit.
- Services make requests to the GLD for equipment requirements unique to their service.
- The GLD probably submits the budget proposal to the CMC through the MND.
- The CMC concentrates on weapons procurement and development, and allocates funds to both. The CMC will determine what proportion of the funding will be derived from the defence budget, the State Council, or other budgetary accounts. For example, weapons and equipment development projects could be funded from budgets assigned to the Commission of Science, Technology and Industry for National Defence (COSTIND), the State Science and Technology Commission (SSTC), or any defence-related ministry, such as the Ministry of Aerospace Industries.

Within this general process,[10] it appears that the final budget is the result of negotiations between the State Council National Planning Commission's Budget Agency, the Ministry of Finance and the party Central Military Commission. The GSD approves or amends funds for operations and maintenance and for equipment. The Services (Air Force, Ground Forces, Navy, and Second Artillery – nuclear forces) and COSTIND present their budget requests through the GSD to the CMC.

The CMC consults with the State Council and obtains a Five-Year Defence Budget Plan. For the annual submission, the CMC makes the final decision on how funds are to be allocated throughout the defence establishment.

The People's Liberation Army Air Force (PLAAF) is believed to have three kinds of budgetary allocations from the GLD that almost certainly reflect the allocations received by other Services. The first is 'fenced' monies designated for the purchase of specific items such as aircraft. The second allocation is 'constrained' funds designated by categories within which the PLAAF has some degree of freedom of choice. The third category is 'discretionary' funds over which there are no restrictions. Under the limitations assigned by the GLD, funds are passed down to Air Force units based on PLAAF budgets and plans.[11] It is likely that all Services and the military region commands operate their budgets within similar constraints and processes.

Prior to analysing what Chinese military expenditures might be, it is essential to look first at the defence allocations found in Beijing's annual budget reports. While sophisticated methodologies have been developed that attempt accurately to measure many aspects of China's economy, each of these methodologies creates problems of its own. Furthermore, there is no commonly agreed interpretation of what these methodologies actually measure.

Since 1980, the Central Intelligence Agency (CIA) has determined that, defined in the manner of the US defence budget, China's military expenditures are approximately twice the published amount.[12] The significance of doubling the official budget allocation, beyond providing a crude sense of how large China's military expenditures might be, is unclear. Determining a US dollar value for any component of the budget in an effort to establish some real value for the allocations faces extremely difficult 'roadblocks', as the CIA readily admits.[13] For example, conversion at the then current official exchange rate of Renminbi – RMB5.8 to the dollar provided a 1993 defence budget value of US$7.4 billion. Conversion at the approximate 'swap rate' of RMB8.8 provided a dollar value of US$4.91 billion. Converting the 1994 defence budget allocation at the new 'market' rate of RMB8.7 provides a dollar value of US$5.97 billion. Thus, in a dollar value determined only by the swap and market rates between 1993 and 1994, the defence budget increased by approximately US$1 billion, or 20 per cent. Quite clearly, conversion to US dollars at the exchange rate has only limited value and does not illuminate what doubling defence budget expenditures means in terms of sustaining and modernizing a force structure as large as the PLA's. Furthermore, in the years covered

Table 4.1 State budget (billion RMB)

	Defence			*Culture/health education/S&T*		
		% change	*% state budget*		*% change*	*% state budget*
1978	16.8		n.a	n.a		
1979	20.2	20.2	n.a	n.a	–	
1980	19.38	–4.1	16.0	15.63	–	12.88
1981	16.8	–13.3	15.1	17.14	9.7	15.34
1982	17.87	6.4	15.5	19.7	14.9	17.08
1983	17.64	–1.3	13.97	20.4	3.5	16.16
1984	18.07	2.4	11.76	26.34	29.1	17.14
1985	19.15	6.0	10.49	31.72	20.4	17.37
1986	20.13	5.1	8.79	38.0	20.0	16.58
1987	20.97	4.2	8.64	40.56	6.7	16.71
1988	21.8	4.0	8.17	47.9	18.1	17.95
1989	25.1	12.6	8.56	51.39	7.3	17.53
1990	29.03	15.6	8.98	61.6	19.9	19.06
1991	33.03	13.8	8.7	69.9	13.5	18.42
1992	37.78	14.4	8.53	78.95	12.9	17.83
1993	43.2	14.3	7.52	96.0	21.6	16.7
1994	52.0	20.4	8.57	113.3	18.0	18.67
	Increase 1985–94 = 171.5%			Increase 1985–94 = 257.2%		

by Table 4.1, the official rate of exchange in US dollars has gone from RMB1.42 to RMB5.8 in 1993, to a daily adjusted market rate hovering around RMB8.7 in March 1994. For this reason, Table 4.1 does not include any conversion to US dollars.

The principal complaint from the PLA over the years since 1984 is that annual defence budget increases were so small they did not keep pace with the inflation generated by China's booming economy. Thus, not only did the PLA fail to benefit as the economy grew, defence expenditures actually shrank when adjusted for inflation.[14] According to Chinese government reports, during the 1980s the military budget declined at an average annual rate of 5.8 per cent in real economic terms as inflation cut into what small increases were provided.[15] The Central Intelligence Agency estimates that, when adjusted for inflation, the defence budget shrank 21 per cent from 1984 to 1988, but grew 22 per cent between 1988 and 1992.[16] In July 1991, the CIA reported that China's rate of inflation reached 27 per cent in the first quarter of 1989, with an annualized rate of 18 per cent. China's defence budget increased by 12.6 per cent in 1989. The annual inflation rate in 1990 was 16.0 per cent,[17] while the defence budget increased 15.6 per cent. In June 1993, the urban cost of living index was some 21.6 per cent higher than the previous June.[18] The defence budget increased by some 12.5 per cent. Thus, the PLA complaint that inflation eroded defence budget increases is supported by CIA analyses.

State budget[19] reports clearly demonstrate that the defence allocations did not keep pace with the rate of inflation. It is difficult, however, to determine precisely how China's general inflation rate effects the PLA as a whole. Weapons and equipment are procured as part of the central planning process. Therefore they will be sold to the PLA at highly subsidized prices. Furthermore, the inflation rate reported is based upon prices in urban areas and then deflated somewhat because inflation in the cities is considerably higher than inflation in rural areas. Yet many PLA units are deployed in rural areas, where they grow much of their own food. Thus, like many Chinese statistics, the reported inflation rates reflect trends rather than precise numerical indices. In this sense, the inflation-corrected decline in defence allocations in the state budget represents a trend rather than a precise indicator of the effect of inflation.

Annual budget reports indicate that for the decade 1985–94, in nominal terms the defence budget increased 171.5 per cent, while allocations for 'culture, education, public health work and science and technology' increased some 257.2 per cent. While any real comparison is troublesome because allocations within such a broad definition as culture, health, education, and S&T are impossible to determine, it is quite evident that within the parameters set by the State Budget procedures, allocation to public welfare did not suffer as much over the decade as defence. Indeed, in 1980, Renminbi or yuan allocations to defence were 24 per cent greater than to public welfare, while by 1994 the public welfare share exceeded the defence allocation by more than 100 per cent. Thus, within China's budget parameters, the balance between defence and public welfare has not only been sustained, allocations for health, culture, education, and S&T have increased at a faster rate than defence spending. There is the distinct possibility that increases in allocations to these areas may accurately reflect the government's response to services that must be sustained in order to maintain social order and the overall development of China's economy. Once again, however, the Ministry of Finance's method of reporting the national budget obfuscates rather than clarifies, for it must be assumed that the defence establishment and the PLA gain from funds allocated to science and technology.

BEYOND THE DEFENCE BUDGET

Recognizing that allocations to defence reported in the state budget do not adequately reflect China's actual military expenditures, a number of techniques have been used to estimate what that expenditure may be. Initially, I will look at the results of applying those techniques to determine if they reflect the PLA's complaint that defence budget allocations have not kept pace with inflation. Table 4.2 reports the results of attempts to provide a more adequate representation of China's military expenditures. All use US dollar equivalents to express these expenditures.

Converting RMB amounts to US dollars faces considerable difficulty. China's currency was not convertible, and the 1993 official exchange rate of RMB5.8 to the dollar overvalued the yuan. The 1993 'swap rate' of approximately RMB8.8 was closer to the yuan's actual market value but, as I noted earlier, converting the 1993 defence budget at the swap rate provides a dollar value of US$4.96 billion – US$3.0 billion less than conversion at the 1993 official exchange rate. The 1994 defence budget converted at the market rate of RMB8.7 provides a dollar value of US$5.97 billion. The dollar amounts in Table 4.2 range from three to four times greater than conversion at the official exchange rate because they are adjusted by purchasing power parity (PPP) calculations and include estimates of additional funding from within the state budget and from other potential sources of income for the Chinese armed forces.

Because Table 4.2 contains efforts to measure the percentage of China's GNP used for defence purposes, it is necessary to make some observation about the method used by the US Arms Control and Disarmament Agency (ACDA) to determine gross national product (GNP): purchasing power parity (PPP). PPP tries to measure the value of Chinese output in US dollars based upon the purchasing power of the country's own currency, but the methods used have produced widely varying results. The various attempts to measure China's 1988 GNP demonstrate how divergent these results can be. A US Census Bureau PPP study produced a GNP value of US$868 billion, the Penn World Table (Mark 5) study resulted in a value of US$2,530 billion, while the ACDA PPP method provided a GNP value of US$471 billion.[20] Table 4.2 uses a conversion rate of RMB1.293/dollar. These same measurement problems create difficulties in attempting to determine what percentage of China's GNP is applied to defence. Using Beijing's statistical reporting, defence uses about 1.8 per cent of China's GNP; doubling the budget figures raises that number to 3.6 per cent. This may be closer to reality, but how close is difficult to determine.

Table 4.2 Military expenditures (ME),[21] in miilion US$*

	ME		*GNP***		*CGE****	*ME*	*ME*
	Current	*1991$*****	*Current*	*1991$*	*1991$*****	*% GNP*	*% CGE*
1981	34,310	51,220	417,600	623,400	145,600	8.2	35.2
1982	36,720	51,620	482,200	677,800	150,500	7.6	34.3
1983	37,380	50,480	553,200	747,300	166,300	6.8	30.4
1984	38,340	49,580	662,400	856,500	190,300	5.8	26.1
1985	39,720	49,560	773,700	965,300	208,100	5.1	23.8
1986	39,930	48,510	860,400	1,045,000	251,300	4.6	19.3
1987	41,590	48,970	985,300	1,160,000	251,400	4.2	19.5
1988	43,870	49,740	1,139,000	1,291,000	249,300	3.9	20.0
1989	45,240	49,080	1,242,000	1,347,000	257,300	3.6	19.1
1990	50,310	52,330	1,369,000	1,424,000	267,700	3.7	19.5
1991	51,040	51,040	1,528,000	1,528,000	293,400	3.3	17.4

Notes:

* Based on a 1991 estimated purchasing power parity of 1.293 RMB/dollar

** Gross National Product

*** Central Government Expenditures

**** Constant 1991 US$

What are China's military expenditures and how much of its GNP is Beijing willing to apply to defence? The simple fact is that outside the numbers provided by the SSB and the Minister of Finance's annual reports, neither China's military expenditures nor its GNP are known and cannot be determined in a manner that is common to all who ask the question. This is especially important when we recognize that the CIA, ACDA, IISS, SIPRI, and other agencies and research centres consider these questions to be important. A measure of a country's concern over defence matters is the percentage of its GNP it is willing to commit to military expenditures. Developing trends in military expenditures will permit exploration of why military expenditures increase and decrease. Current methodologies evidently do not permit common answers to these questions simply because no satisfactory methodology has been derived that will allow the transformation of weak data into robust numbers. Analysts for IISS, for example, recorded a range of uncertainty ranging from US$11 billion to US$23 billion.

Disregarding the differences in the numbers created by different methodologies, it is also quite evident that the attempts to determine China's military expenditures support the PLA's grievance that defence allocations did not keep pace with the rate of inflation. In the eleven years reported in Table 4.2, which contain the years of most serious inflation, ACDA estimates of China's military expenditures indicate that when measured in constant 1991 US dollars, defence expenditures actually shrank. Measured in current dollars, expenditures over the decade grew by 48.76 per cent. As a percentage of GNP and CGE, ACDA's PPP-adjusted estimated military expenditures shrank by over 50 per cent. Thus, no matter which set of numbers is used, whether official budget allocations or estimates of Beijing's actual military expenditures, the Chinese armed forces experienced a serious decrease in military funding over the years 1982–88. The budget increases of 1989–94 brought the armed forces back to the point where their resources were sufficient to counterbalance or get slightly ahead of inflation. To what extent the increases in real terms enabled the armed forces to make up for the serious reduction in expenditures experienced in the mid- to late 1980s is impossible to determine from the data available.

The official budget, as is well known, does not fully reflect China's military expenditures. There are multiple potential sources for additional defence funding, including allocations from other budgetary accounts. In recent years, however, considerable attention has been paid to sources of revenue for the PLA originating in arms sales and commercial enterprises.

'PLA INCORPORATED': EXTRA-BUDGETARY SOURCES OF REVENUE

Whereas the defence budget process is responsible for major budget items and the manner in which they are to be funded, there is an additional and potentially extensive source of funding stemming from PLA-owned and operated commercial enterprises such as hotels, factories, automobile and truck production, arms sales, farms, mines, airline and shipping services, and many other forms of business, including dealing in currency futures. These enterprises are owned and to a great extent managed and operated at all levels of the armed forces from the three General departments down to the division level and below, including the People's Armed Police forces. Military commerce is now so extensive that the PLA is constructing the Niutianyang economic development zone near Shantou in the prospering coastal region of Guangdong province. The zone encompasses 19 sq. km and will be a completely self-contained urban centre incorporating 'financial, commercial, industrial, tourist and residential blocks'. The development zone will be owned and operated by the Guangzhou Military Region.[22]

PLA-connected businesses are now encouraged to form transnational enterprise groups. The 999 Enterprise Group (999) was formed at the direction of the CMC and the GLD in early 1992. Formed around the Nanfang Pharmaceutical Factory, the 34-member group has fixed assets of 1.6 billion yuan with operations involving medicine, real estate, electronics, motor vehicles, clothing, trust investments, some of which involve joint ventures with foreign firms. By December 1992, the 999 Enterprise Group had established branches in Russia, Germany, Thailand and the United States. With a development strategy extending over ten years, 999 clearly intends to become larger and more international in its market. What is more, the press briefing where 999's achievements and development strategy were proudly announced was attended by such leading military figures as Generals Chi Haotian and Zhang Aiping. The two senior military members of the CMC, Generals Liu Huaqing and Zhang Zhen, sent a congratulatory message to be announced at the news briefing.[23]

Beyond the supervision provided by the GLD's Production Management Department that oversees PLA-run enterprises,[24] it is quite unclear how these businesses are regulated and how the revenues derived from them are allocated. Nor does the GLD have any real knowledge of how many businesses the PLA may be operating. There may be as many as 20,000, but because military units put together unauthorized enterprises, and successful corporations fund spin-offs without clearing them with Beijing, there is no actual number the GLD can declare with confidence.[25] Nor is it certain how much money these enterprises make. Some analysts in Beijing speculate that the total revenue could match the annual defence budget – 42.5 billion yuan in 1993.[26] The CIA recognizes that the income derived from these commercial enterprises is unknown and provides the Hong Kong press report that the amount could have been 30 billion yuan in 1992.[27]

Major General Zhu Zuoman, president of GLD-owned *Xinxing*, says that the corporation produced income of almost 7 billion yuan in 1992, with a profit of around 700 million yuan.[28] *Xinxing* corporation may well be the largest of all PLA conglomerates – perhaps 20 per cent of all the enterprises owned by the General Logistics Department. Established by the GLD in 1984 to make better use of under-utilized resources, by 1987 *Xinxing* owned in excess of 3,000 factories, 2,000 farms, and 8,000 enterprises that included universities, scientific research centres, and hospitals. *Xinxing's* additional properties included 'dozens of mines and 16 large horse ranches'.[29] The most recent development within 'PLA incorporated' has been the evolution of 'enterprise groups'. These groups serve two purposes. First, many commercial enterprises are undertaken by lower level military units and are too small to be economically efficient and sound. By merging these enterprises, the GLD hopes to achieve economies of scale and thereby make them profitable. Other enterprise groups are simply the creation of conglomerates like the GLD's *Xinxing* Corporation that began as a trading company and in 1991 was transformed into a conglomerate directing a wide variety of enterprises.[30]

The GLD's Whole Army Production and Operation Conference, held in February 1993, reported that the PLA set up twenty enterprise groups in 1992. The conference also noted that military enterprises are involved in joint ventures with foreign firms. While not providing detailed information, the session reported that there was an increase of 200 per cent in the amount of foreign capital utilized by PLA joint ventures in 1991, and a 230 per cent increase in 1992.[31]

There can be no doubt that PLA commercial activities are extensive and continue to grow. General Zhang Wannian, CMC member and chief of the PLA General Staff Department, was quite forthright in December 1993 when he observed that 'deficiencies' in the defence budget had been made up by military commercial enterprises. He reported that revenues from the PLA's 'production and business' had enabled the armed forces to engage in 'normal training and operations against war' and to improve the armed forces' living conditions.[32] General Zhang made it equally clear that there are many serious problems connected with these enterprises. A central issue is corruption and mismanagement of military enterprises that have seriously eroded the public image of the PLA.

Four years earlier, the then-Director of the General Logistics Department, General Zhao Nanqi, at the December 1989 All-Army Logistics Work Conference outlined a five-point guideline that called for 're-organizing and developing military-run enterprises to supplement the military's income and increase its power to buy equipment'.[33] While details of the guidelines were not provided, Zhao Nanqi's concern over corruption and the misuse of enterprise funds was made clear. He warned against buying elaborate houses, expensive office equipment, setting up illicit bank accounts, buying cars in excess of those authorized, dining, travelling and providing gifts at public expense, misappropriating and withholding allowances and materials allotted for military purposes.[34] The problems Zhao Nanqi outlined in 1989 were accentuated by the fact that the brutally suppressed demonstrations in Tiananmen Square that spring were in part brought about by the reaction against the corruption that infects all levels of official life, both civilian and military. Evidently these problems have yet to be resolved and have become even more severe over the past few years. As early as 1988, in an effort to separate the armed forces image from its commercial activity and the shadow of corruption, the CMC had directed that while the PLA could own enterprises, it could not manage them; that 'cadres' managing and operating these businesses should not remain in the armed forces and should not be paid from the military budget.[35] This attempt to separate the armed forces has not yet been implemented and there are continuing reports of attempts to separate PLA-run enterprises from the military by separating ownership from management. It is unlikely that these efforts will succeed in the immediate future. The most recent report states that Army Enterprise Bureaux will be established in each of the seven MRs to consolidate and curtail commercial ventures run by PLA units. The same report states that the CMC has agreed to give up many military enterprises, including mines and transportation businesses – but only after compensation from the government.[36]

Arms sales

Arms sales have also been the focus of considerable speculation as a source of extra-budgetary funding. As a function of the Iran–Iraq war, China joined the West and the Soviet Union as one of the world's largest arms' vendors. (See Chapter 6.)

Reviewing the data on an annual basis, it is evident that the estimated value of China's arms sales is quite small; significantly less than the value of arms and equipment sold by the USA or the USSR/Russia. Over the 8-year period, China's sales averaged little more than US$1.5 billion a year. Even if the entire amount were profit and all the profits were turned over to the PLA, the addition to the PLA's coffers is minimal – even accepting that every penny (fen?) helps.

There are several problems that must be raised before attempting to determine the extent to which China's arms sales were a source of extra-budgetary financing for the PLA. First, the Chinese government itself notes that one third of state-owned enterprises are operating at a loss and that only one third are breaking even. Arms industries are state-owned. Given the dual-pricing system followed in China, it is more than likely that export production does not receive state-subsidized materials. Are these state-owned factories operating efficiently and do they make much profit? If they make any profit, they are among the few state enterprises that do.

Second, even though we know that several major arms purveyors are subsidiaries of the Services or the PLA General Departments, do we know that profits are returned to PLA? It is quite probable that profits are returned to the owning corporation, whether to ministry, military service, or one of the three General Departments that directs/owns the corporation. After the corporation is repaid and the business costs reimbursed, how much profit is returned to the armed forces? Is the profit returned to the individual service or to the General Staff, Political or Logistics Department as part of the funding for the five-year budget plan, or is it used to supplement the annual budget submission?

The information currently available does not permit any firm conclusion about the amount of extra-budgetary sources of funding derived from China's arms sales. Certainly there must be some financial gain, but at most it is small and does not represent a major source of funding. What it may do is to provide the CMC with some convertible currency to apply to the acquisition of foreign technology or, in a very limited way, advanced weapons and equipment. Nonetheless, the amount of funding subsidized by arms sales has to be small because the gross sales are small even before profit is estimated.

Defence conversion

It is also unclear how much of the revenue generated by civil production from China's defence plants is transferred to the PLA (see Chapter 5). Chen Dazhi, Director of COSTIND's Planning Department, reported that these industries will invest 'at least' 6 billion yuan (US$1.03 billion) both to modernize China's defence plants and convert them to civil production. By 1995, more than 80 per cent of the defence plants will be able to produce civilian goods.

Given the number of PLA-owned corporations selling civilian as well as military goods abroad, it is reasonable to assume that some of the profits return to the Services and the PLA General Departments.

These PLA-owned trading firms will return part of their profits to the Services and the three General Departments that operate them. Once again, however, how much of the revenue generated is profit and how much of that is provided as an extra-budgetary source of funds for the PLA is not known. Nor should it be assumed that all defence-related conglomerates are moneymakers.

The MR-level and below: enterprises and farms

Some 30 per cent of MR and military unit budgets must be provided by the MRs and units themselves. Military regions, group army level and lower units generate revenue from farms and enterprises run by soldiers and/or their dependents, and from investments. Part of the profit from MR enterprises must be returned to the GLD in Beijing. Some of the revenue from military unit enterprises and farms is passed to the MR headquarters, but most profit is retained by the individual units. In addition to the revenue gained from selling agricultural products, PLA units from all the Services and the MRs use their farms' products to improve the diets of soldiers and their families. Some of the revenue is also used to improve the living conditions of soldiers and dependents. Interestingly, it appears that funds returned by MRs located in those parts of China undergoing rapid economic growth, such as the Guangzhou MR, are used to improve the living conditions of soldiers in remote border areas such as Xinjiang and Tibet, where the opportunity for moneymaking enterprises is considerably less than in coastal China. While such a process could work well, the significant and growing complaint from the GLD about corruption in PLA enterprises and its insistence that all PLA-run enterprises follow the strict regulations the GLD has promulgated, indicates that there is a major problem. Specifically, MR and PLA unit enterprises are deceiving Beijing, making it impossible for the GLD to recover its due.

There is even less information about revenue derived from farms, enterprises, factories, and investments operated by the many lower levels of the PLA than there is about arms and equipment sales, or commercial enterprises owned and operated at the highest levels of the military hierarchy. The number of such revenue producing activities must, however, be huge. A report from the Liaoning provincial Military

District (MD) provides a clear image of the range and potential for PLA enterprises at the military district level.[37] In April 1993, the Liaoning MD's development plan for the year was designed to provide an annual average increase in profits of 20–25 per cent through expanding the MD enterprises' penetration of the province's growing market economy. Since 1985, the MD had established more than 130 military-run enterprises – food, petrochemical, catering, and service trades employing in excess of 5,000 staff and workers. Output was 729 million yuan, providing 116 million yuan profit and 21 million yuan in taxes. Revenues from the enterprises contributed 78 million yuan for 'building' the army, militia, and reserves. Responding to the GLD's guidance that managing military affairs be separated from enterprise management, the development strategy for 1993 included the provision that the MD's enterprises would be separated from its military responsibilities. Presumably, the enterprises would be managed and staffed by military dependents or by officers who retire. There is no reason to consider the Liaoning MD unique in its approach to enhancing its sources of revenue, and the provincial military districts of Guangdong and Fujian at the epicentre of China's booming growth triangle with Taiwan and Hong Kong must be developing enterprises and profit at an even faster pace.

CONCLUSIONS AND SPECULATIONS

The purpose behind determining military expenditures is to discover what defence burden a state is willing to bear. Changes in defence burden are perceived as reflections of altered threat perception or that a state is engaged in a potentially aggressive military build-up.

While ACDA's estimated 1991 military expenditure of US$51+ billion places Beijing in third place behind the United States with a US$280+ billion budget and the former USSR with an estimated US$260 billion, other rankings have far different outcomes. ACDA estimates of military expenditures as a percentage of GNP place China 64th out of 142 countries. But the simple fact is that the magnitude of China's estimated military expenditures places Beijing among the world's leading military powers. On all other measures, China ranks among the developing countries of the world.

What distinguishes the Chinese military from other developing countries is its size and nuclear weapons. In all other respects, the Chinese armed forces reflect the poverty of the PRC's industrial and S&T base. Indeed, the technological sophistication of the PLA's conventional weapons and equipment lag far behind most of its neighbours.

The weapons and equipment of the Chinese armed forces demonstrate the neglect of decades. There are always rumours of a new and sophisticated 'something' about to emerge from Xi'an, Chengdu, Shenyang, Wuhan, Shanghai, or the deserts of Lop Nor. When 'it' does appear, not only is it decades behind similar weapons or equipment designed and produced by advanced industrialized societies, but it also takes a decade or more to put the system into series production. No matter what estimates one accepts as accurately reflecting China's military expenditures, these expenditures have not compensated for thirty years of neglect. Without a massive technological transfer similar to that provided by the USSR between 1954 and 1960, the Chinese armed forces are condemned to prolonged and indefinite modernization.[38]

While estimating China's military expenditures is a necessary exercise, and will remain so until the PRC's defence budget more closely reflects Beijing's actual military expenditures, it is important to avoid excessive interpretation of what these estimates may imply. Military expenditures must be understood in relation to the pattern of China's military acquisitions, national military strategy, doctrine, and training before any meaningful inferences can be drawn. When estimates of military expenditures are subject to extremely large margins of error because of the weak database upon which the requisite methodologies must be based, analysts must acknowledge the tentative nature of their conclusions. Thus, the first two major increases of 1989 and 1990 were designed to improve pay and other benefits that had been severely eroded by the inflation of the 1980s. Increases since then reflect continuing attempts to improve pay and other benefits, such as housing, but include the modernization of weapons and equipment. Nonetheless, military modernization remains a very low priority and evidence indicates that overall force modernization will occur as a product of industrial and S&T modernization. Large-scale modernization of weapons and equipment, as opposed to one-off prototypes shown at international arms exhibitions, cannot occur until well into the twenty-first century. The principle limitations creating this condition are the technological backwardness of China's industrial base and S&T infrastructure combined with the long development times involved in evolution and production of advanced weapons systems. Without clear information on procurement and research, development, test and evaluation expenditures there is literally no way to determine what Beijing's precise weapons and equipment modernization goals and priorities are. Estimating these expenditures from the database now accessible to analysts is virtually impossible.

From all this I conclude that estimating China's military expenditures as an exercise separate from a careful analysis of changes in national military strategy, doctrine, and concepts of operations is of only limited value in understanding Beijing's defence priorities and capabilities. The margin of error remains just too great for any meaningful conclusions. Judicious analysis combining such estimates with other indicators of defence policy will provide more fruitful findings.

NOTES

1 Japanese Foreign Minister Hata, for example, explicitly raised China's defence budget increases in his January 1994 meeting with President Jiang Zemin. Takahiko Yokoyama, 'Japanese–China Relations May Step on Thin Ice', *Asahi Shimbun*, 10 January 1994, p. 2.
2 For example, in April 1993, Senate Majority Leader George Mitchell warned that China's rising defence budgets and acquisition of sophisticated military technology raises the long-term security concerns of the United States and its Asian allies. See Robert G. Sutter's discussion in *China As A Security Concern in Asia: Perceptions, Assessment, and U.S. Options*, CRS Report for Congress 94–32 S, Congressional Research Service, The Library of Congress, 5 January 1994, p. CRS-2.
3 Unless otherwise specified, all numbers of Chinese military personnel, weapons and equipment are taken from *The Military Balance*, London: The International Institute for Strategic Studies, Fall 1993.
4 Ibid.
5 Dr Alfred D. Wilhelm of the Atlantic Council is the originator of this phrase.
6 Deng Xiaoping, 'Speech at an Enlarged Meeting of the Military Commission of the Party Central Committee', 14 July 1975, *Deng Xiaoping Wenxuan* (Selected Works of Deng Xiaoping), Beijing, 1 July 1983, in Joint Publications Research Service (hereafter JPRS), *China Report*, no. 468, 31 October 1983, p. 19.
7 Reported by *Xinhua*, Beijing, 27 March 1991, in Foreign Broadcast Information Service; *Daily Report: China*, (henceforth FBIS-CHI), 28 March 1991, p. 19.
8 See James Harris *et al.*, 'Interpreting Trends in Chinese Defense', in Joint Economic Committee, Congress of the United States, *China's Economic Dilemmas in the 1990s: The Problems of Reforms, Modernization, and Interdependence*, US Government Printing Office, Washington DC: April 1991, p. 679.
9 Kenneth W. Allen, Major, USAF, *People's Republic of China People's Liberation Army Air Force*, Washington DC, Defense Intelligence Agency, DIC-1300–445–91, May 1991, p. F-20.
10 Information gleaned from a number of sources and interviews.
11 Ibid.
12 National Foreign Assessment Center, *Chinese Defense Spending, 1965–79: A Research Paper*, Central Intelligence Agency, Washington DC, SR 80–10091, July 1980, p. 1; Directorate of Intelligence, *The Chinese*

Economy in 1991 and 1992: Pressure To Revisit Reform Mounts, Central Intelligence Agency, Washington DC: EA 92–10029, August 1992, p. 11.

13 This term is used in, Directorate of Intelligence, *The Chinese Economy in 1990 and 1991: Uncertain Recovery*, Central Intelligence Agency, Washington DC: EA 91–10022, July 1991, p. 14.

14 There have been numerous discussions focused on the effects of inflation on the defence budget in recent years, among these are: Liu Tianyi, 'Limited Military Expenditures Make It Necessary to Practice Economy – Zhao Nanqi Answers Questions on the State of the 1987 National Defence Budget', *Jiefangjun Bao*, 31 March 1987, in FBIS-CHI, 10 April 1987, pp. K25–26; Huang Kangsheng, 'The Road of Moderate Development Should be Taken for National Defense Construction – Interview with Wang Qikun, President of the Academy of Military Economics', *Renmin Ribao*, Overseas Edition, 29 June 1990, in FBIS-CHI, 10 July 1990, pp. 30–32; and Chen Bingfu, 'An Economic Analysis of Changes in Chinese Military Expenditure in the Last Ten Years', *Jingji Yanjiu*, no. 6, 20 June 1990, in FBIS-CHI, 6 August 1990, pp. 30–35.

15 James Harris *et al.*, 'Interpreting Trends in Chinese Defence Spending', fn. 2., p. 677.

16 Directorate of Intelligence, *The Chinese Economy in 1991 and 1992*, p. 11.

17 Directorate of Intelligence, *The Chinese Economy in 1990 and 1991*, pp. 1–2.

18 Directorate of Intelligence, *China's Economy in 1992 and 1993: Grappling With the Risks of Rapid Growth*, The Central Intelligence Agency, Washington DC: EA 93–100016, August 1993, p. 18.

19 All budget figures are taken from the Minister of Finance's annual reports published in FBIS-CHI for the appropriate years.

20 For a discussion of these PPP applications, see Directorate of Intelligence, *The Chinese Economy in 1990 and 1991*, Appendix A, pp. 15–16.

21 United States Arms Control and Disarmament Agency, *World Military Expenditures and Arms Transfers 1991–1992*, Table I, p. 58, Washington DC: USGPO, March, 1994. In their 'Statistical Notes', p. 151, ACDA analysts are careful to warn the reader that the exceptional difficulties in determining yuan costs of Chinese forces, weapons, and programmes mean that all data related to Chinese military expenditures have to be seen as encompassing a wide margin of error.

22 Beijing, *Xinhua*, 12 February 1993, in FBIS-CHI, 12 February 1993, p. 17.

23 Beijing, Zhongguo Xinwen She, 29 November 1992, in FBIS-CHI, 9 December 1992, pp. 33–34.

24 Wang Yihua, 'What Is Seen Regarding Troops' Production and Economic Activities and Thoughts Provoked Thereby', *Jiefangjun Bao*, 13 December 1992, in FBIS-CHI, 13 January 1992, p. 42.

25 Tai Ming Cheung, 'Serve the People', *Far Eastern Economic Review*, Vol. 156, no. 41, 14 October 1993, p. 64.

26 Ibid., p. 65.

27 Directorate of Intelligence, *China's Economy in 1992 and 1993*, p. 9.

28 Ibid., p. 65.

29 *China Daily*, Business Weekly Supplement, 8 April 1987, in FBIS-CHI, 14 April 1987, p. k-31.

30 Ibid., p. 65.

31 Zhou Tao and Ma Chunlin, 'Whole Army Production and Operations Hit Record Highs', *Jiefangjun Bao*, 18 February 1993, in FBIS-CHI, 2 March 1993, p.27.
32 Guo Jia, 'Zhang Wannian Calls on Various Leading Military Organs to Take Overall Situation Into Account, Set Example of Observing Discipline', *Renmin Ribao*, December 1993; in FBIS-CHI, 21 December 1993, p. 32.
33 Beijing, *Xinhua*, Domestic Service, 25 December 1989, in FBIS-CHI, 29 December 1989, pp. 22–4.
34 Ibid.
35 Ibid., p. 41.
36 Willy Wo-lap Lam (no title), *The South China Morning Post*, 9 December, 1993, in FBIS-CHI, 9 December 1993, pp. 20–21.
37 Shenyang, *Liaoning Ribao*, 4 April 1993, in FBIS-CHI, 27 April 1993, p. 31.
38 See, for example, the analysis presented by Robert J. Skebo, Gregory K.S. Man and George H. Stevens (all serving military officers from DIA), 'Chinese Military Capabilities: Problems and Prospects', in *China's Economic Dilemmas in the 1990s*, pp. 663–75.

5 Economic reform and defence industries in China

Arthur S. Ding

The Third Plenum of the Chinese Communist Party's (CCP's) Eleventh Central Committee in 1978 was an important milestone in the development of the Chinese economy. The introduction of the household contract system in the countryside loosened the grip of the planned economy. Then the reforms were extended to the urban areas and the regime had to use complex market instruments, combined with administrative measures, to carry out overall economic guidance. The marketization process has inevitably experienced ups and downs, but steady progress has been made in general.

However, this progress from an initial relaxation of controls towards a market economy has had a considerable impact on individual elements of the economy, particularly the state-run enterprises. China's defence industries, as a part of the state sector, have shared this impact. But because of their special responsibilities – in terms of research and development (R&D) and the production of weapons necessary for national defence – the experience of these industries has been unique and they have suffered different problems.

This chapter will discuss the impact of China's economic reforms on the country's defence industries. It will begin with an outline of the reform process, and go on to discuss the impact of the reforms on defence industries, how these industries have adjusted their production and product structure, and the regime's evaluation of this restructuring. Finally, the author will attempt to outline the challenges that these industries are likely to face and their prospects for the future.

CHINA'S ECONOMIC REFORM PROGRAMME

China's economic reform programme, which was started over a decade ago, has been wide-ranging in its content and scope.[1] It has included reform of the price system, the ownership system, and the management mechanisms of enterprises, all of which are aspects of the planning system.

Soon after it was founded in 1949, the People's Republic of China (PRC) adopted a system of centralized economic planning which completely discounted the role played by the price mechanism. To regulate economic activities, the regime relied entirely on administrative means, while the state monopolized the means of production, controlled the supply of consumer goods, and fixed the prices of products.

The prices fixed by the state reflected neither the supply and demand situation nor the real value of the products. What is more, since the power to fix prices was concentrated at a high level in the administration, localities and individual enterprises were deprived of incentives and flexibility, preventing them from exercising their initiative. Prices were fixed indefinitely and their failure to reflect market conditions prevented them from regulating production and consumption.

Price reform in the PRC began with the large increase in the purchasing price of agricultural products introduced in 1979. This aspect of reform has been carried out according to the principle of 'combining devolution [of pricing powers] with adjustment [of prices], and progressing step-by-step'. As far as possible, price stability has been maintained so as to avoid social unrest.

The devolution of pricing power expanded the role of localities and key cities in fixing prices, and allowed enterprises to fix or adjust prices in line with costs and the supply and demand situation, offering them added incentives and giving them increased scope to exercise their initiative. This represented a major transformation of the PRC's price control system.

Permitting the planned adjustment of prices for commodities and labour services meant allowing prices to rise and fall, rather than interpreting price stability as a freeze on prices, as had been the case in the past. On this basis, the PRC has gradually increased the role of the market mechanism in price adjustment, and reduced the scope of price fixing by the state. This has involved the creation of a three-tier price system consisting of a state fixed price, a floating price, and a free market price, while the power to fix prices is shared among the central government, the localities, and the enterprises themselves. Although this has had the effect of making prices more realistic, and less distorted, it has created new problems for China's defence industries.

Where ownership is concerned, the reforms have created a situation in which several different forms of ownership exist simultaneously. During the Mao era, the ownership system in the PRC underwent three major changes, culminating in a unified system of state ownership. However, this system destroyed the vitality of the economy by over-emphasizing heavy industry, particularly large and medium-sized industrial enterprises, which ate up a disproportionate amount of resources.

The current reforms have involved the establishment of a multiple ownership structure in which public ownership plays the leading role. In theory, 'ownership by the whole people' is the guiding force of socialism, collective ownership is an important element of the socialist economy, while individual ownership, private companies, and foreign invested enterprises play a necessary supplementary role. In practice, the regime permits individual economic activities in urban and rural areas, as well as allowing the establishment of private firms using hired labour. Chinese firms may also form joint ventures with foreign businesses to attract foreign investment and import advanced technology and management methods. In line with the belief that 'ownership by the whole people' is the guiding force in the socialist economy, the PRC gives priority to the state-owned sector in terms of investment in capital construction, raw materials, and labour.

Despite the fact that the regime's policy toward the non-state sector tends to be variable and contradictory, this sector has developed extremely rapidly and is thriving. According to PRC statistics, the gross output value of the state-owned industrial sector as a percentage of total industrial output decreased from 75.97 per cent in 1980 to 56.8 per cent in 1988, and it continued to decline in the 1990s. The collective industrial sector, on the other hand, experienced a rise in output value over the same period from 23.54 per cent to 36.15 per cent of the total, while the share of the individual sector rose from 0.02 per cent to 4.34 per cent and that of other sectors rose from 0.47 per cent to 2.71 per cent. Generally speaking, individual, private, collective, and foreign invested enterprises are playing an increasingly important role in the PRC economy.

What is more, the non-state sector, which has more autonomy and is less restricted by state planning and price controls, is far more efficient than the state-owned sector. Comparatively speaking, the private sector performs better than the individual sector, the individual sector performs better than the collective sector, which in turn outperforms the state-owned enterprises, while in terms of size, small enterprises are much more successful than large and medium-sized ones. Failing to thrive on its own, the state-owned sector must rely on administrative means to survive competition from the non-socialist economy, and it is questionable how long it will be able to do this.

The 'Decision on the Reform of the Economic Structure', passed by the Third Plenum of the CCP's Twelfth Central Committee, contained measures specifically designed to reinvigorate state-owned enterprises and formed the crux of the PRC's urban economic reforms. Changes in the management mechanisms of state-owned enterprises may be grouped into two categories: separation of management from ownership, and an adjustment of the relationship between enterprises and the state. The separation of management and ownership has taken the form of transferring management responsibilities to individuals or groups by means of legal contracts or agreements. Different versions of this include the leasing system and the property management responsibility system, while others include the 'factory boss (manager) responsibility system' and the 'tenure goal responsibility system'. These were all attempts to separate ownership from management without fundamentally changing the state ownership system.

Changes in enterprise–state relations gave enterprises more autonomy. In addition to the above measures to increase the powers of enterprise managers, the regime also substituted tax payments for the previous system of handing over profits to the state. In theory, this gave enterprises more financial autonomy, enabling them to carry out reforms and develop their operations. Related to the increase in enterprises' financial autonomy was an increase in their rights in the market. At the same time, the regime gradually reduced the number of products and materials controlled by the mandatory state plan.

However, there has been only a limited improvement in the management of large state enterprises, and in recent years the regime has issued a whole series of regulations aimed at further reforming their management mechanisms so as to halt their decline. Despite these efforts, large state enterprises remain in crisis, and defence industries, most of which fall into this category, are no exception.

THE PRC'S DEFENCE INDUSTRIES

Defence industries[2] are defined as economic entities which conduct research, development, and manufacture of weapons. In the early years of the regime, the PRC established a total of seven engineering industry ministries to take charge of military weapons research and manufacture, as well as two bodies responsible for overall planning and decision making: the National Defence Industry Office (Guofang gongye bangongshi) and the National Defence Science and Technology Commission (Guofang keji weiyuanhui). In 1982, these agencies underwent reorganization and merger, though their functions remained unchanged. The reorganization involved the merger of the National Defence

Industry Office and the National Defence Science and Technology Commission, together with the Science and Technology Equipment Commission (keji zhuangbei weiyuanhui), to form the Commission of Science, Technology, and Industry for National Defence,[3] thus gathering responsibility for research and development (R&D) and production in one body. The seven ministries were then transferred from military supervision to become part of the State Council and renamed as follows: the Ministry of Machine Building, Ministry of Nuclear Industry, Ministry of Aviation Industry, Ministry of Electronics Industry, Ministry of Ordnance Industry, Ministry of Shipbuilding Industry, and the Ministry of Astronautics Industry.[4]

This change to civilian control marked the beginning of the conversion to civilian production (*jun chuan min*) among the PRC's defence industries. Deng Xiaoping had mentioned the need to combine military and civilian production in the PRC's defence industries as early as July 1978, and a policy was subsequently formulated which called for military plants to subsidize the production of weapons through production for the civilian market.[5] If the seven ministries had remained under the control of the CCP's Central Military Commission, it would have been extremely difficult to carry out such a shift in production. It would have been impossible to reinvigorate China's defence industries, let alone attract foreign capital and technology transfers or enter into any form of cooperation. Transferring the PRC's defence industries to the normal civilian administration created the right environment for their recovery and provided an opportunity for them to develop their productivity.

The PRC's defence industries were based on the Soviet model in that the government invested large quantities of manpower, capital, and materials to establish a military industrial system that was completely independent from the civilian sector. No regard was given to the needs of the non-military sector or the relationship between defence and civilian industries. What is more, each industry ministry had its own large and independent research, design and production capability.

In terms of management, the Soviet-style defence industries were very similar to state-owned civilian enterprises. Management authority was centralized in the hands of the supervising ministries, and individual enterprises were deprived of any rights over personnel, organization, product development, or the allocation of capital. The management structure was closed and hierarchical, with authority emanating from the defence industry ministries. Under this system, enterprises under different ministries were prevented from forging any links with each other. There was heavy reliance on compulsory planning, and all raw materials, capital, and products were allocated uniformly according to the plan. Since enterprises had no autonomy and were merely agencies of their respective ministries, their only function was to carry out production in line with the ministries' orders. The chain of command operated entirely from the top down and there was no opportunity for feedback from below.[6]

By the early 1980s, the PRC authorities were well aware of the drawbacks of this system and sought to cure them through the application of the economic reform measures and administrative changes described above, and through promoting a conversion to civilian production. One of the drawbacks was the existence of completely independent and self-sufficient systems, big and small, and a lack of division of labour. Inefficient military industrial enterprises were hampering progress in defence technology, and they were also imposing an unbearably heavy burden on the government. Local governments found the burden particularly onerous, as they were responsible for providing the enterprises with backup services and underwriting them financially, while military plants brought them little return in terms of taxes.[7]

THE IMPACT OF THE ECONOMIC REFORMS

The PRC's defence industries have been greatly affected by the economic reform programme in 1978. For a start, defence enterprises began to suffer from surplus production capacity.[8] The reforms led to a reallocation of resources and the defence budget was cut. This situation was exacerbated as the PRC set about creating a peaceful international environment for reform and improved relations with the Soviet Union. All these factors contributed to a sharp drop in orders for arms and military equipment which left military plants lying idle. At this stage, the PRC authorities were only just beginning to re-examine the structure of the defence industries and the relevant government policies,[9] and had yet to come up with any effective solutions to their problems. Thus, the very survival of the military industrial enterprises was under threat.

With order books empty and the government yet to come up with a feasible policy, individual enterprises began to seek their own solutions to the problem.[10] But the long years of centralized economic planning under which enterprises were no more than an extension of the bureaucracy, responsible solely for production, had deprived these plants of the ability to think up suitable products to fill their surplus production capacity. And apart from that, they lacked access to market information.

PRC publications of the time contain vivid descriptions of the plight of the military plants. As one writer describes, 'at that time, it was basically a case of doing anything one could, some nuclear arms factories produced soft drinks, air fields produced chicken coops, mirrors, dressing tables, etc'.[11] Articles such as these point out that the choice of product depended entirely on short-term market demand. Some plants were able to produce slightly more sophisticated products, such as bicycles and electric fans, but excessive costs made them uncompetitive.[12] The PRC authorities were well aware of this situation and in 1983 began to take steps to resolve it. In that year, the top leadership endorsed the need to combine military and civilian production, making the shift to civilian production a long-term policy rather than a temporary expedient. Starting in July 1986, more specific measures were adopted. Certain civilian products manufactured by military plants were included in the central plan, military enterprises were urged to cooperate and introduce division of labour, they were allowed to conduct major development and production projects under contract, and the government made provisions to subsidize their technological upgrading.

Despite all these efforts, the PRC's defence industries still had an enormous surplus production capacity, with between one- and two-thirds of plant lying idle.[13] The worst afflicted were the 'third line' enterprises which were built in the interior provinces in the 1960s. Also, large-scale plants – like those producing conventional weapons – and those employing more sophisticated equipment – such as nuclear weapons' plants – were more likely to have equipment lying idle on account of the difficulty of converting them to civilian production.[14]

The second impact of the reforms was increased financial stringency.[15] The conversion to civilian production was all very well in theory, but it required a quite considerable investment of capital if it was to be successful. The reforms, however, meant that funding was even more scarce. First, this was because the drop in orders for military equipment brought a reduction in direct government investment. Second, the prices paid by the government for military equipment had long been far too low; the more the enterprises produced, the more losses they incurred, so there was no surplus available for reinvestment. Arms and military equipment were not treated as commodities and prices had always been calculated to cover production costs plus a 5 per cent profit. What is more, during the course of the reforms, the government began to free the prices of many products, while labour costs were also rising. There was little scope, however, for adjustment in the prices of military products. These factors reduced the efficiency of the military enterprises still further.

Reports in PRC publications reveal the seriousness of this problem. Because the quantity and quality of raw materials and parts supplied under the central plan were unsatisfactory, military enterprises were forced to buy on the free market, incurring serious losses in the process. In one instance it was reported that an enterprise 'would lose an average of 1,000 yuan on major repairs to one military vehicle, 5,000 yuan on the repair of a piece of engineering machinery, and 50–100,000 yuan on a ship'.[16] The more repairs they carried out, the more losses they incurred. Military-related work had to be subsidized out of the profits of civilian production, depriving the enterprises of funds to develop new civilian products.

Another factor which exacerbated the shortage of funds was the closed and segregated nature of the PRC's administration. This consists of a vertical (*tiao-tiao*) structure extending from the central ministries down to the enterprises under their control, and a horizontal (kuai-kuai) system centred on individual provinces. As mere extensions of the bureaucracy, enterprises belonged either to one system or the other, and each had its own planning controls. The financial aspect of the economic reforms intensified this segregation. Localities handed over a negotiated proportion of their tax revenues to the central government and were allowed to retain the rest to be used for local development. Local governments began to employ administrative means to maximize their tax revenues, including giving the enterprises under their jurisdiction priority in the allocation of capital and raw materials under the plan. Military enterprises, which were part of the vertical system, were deprived of support from the localities.[17]

Lack of funds forced military enterprises to use outdated technology and equipment. Most of the PRC's military industrial plants were built in the 1950s and 1960s and their equipment is in urgent need of renewal. According to one PRC source, in 1990 the equipment in some military industrial enterprises was barely on a par with that used by township enterprises.[18] Deficiencies in equipment naturally affect the quality of products and their ability to compete in the market.

The third impact of the reforms is the depletion of the work force in military enterprises. Under the economic reforms, monetary rewards have largely replaced contribution to society as a measure of achievement. This factor, plus the fact that the very survival of many military enterprises is in question and that the authorities now permit workers to change jobs, has caused an outflow of labour from military enterprises to foreign joint venture companies and private firms. Those workers who remain, particularly the younger ones, go to great lengths to get transferred to civilian production departments. Those departments engaged in military production are left with an ageing workforce and little chance of finding replacements, and this is causing a crisis in the development of defence technology.

The quality of the workforce is also deteriorating as a side effect of the PRC's social system. The strict controls on movement enforced in the PRC over the past forty years (including state allocation of employment and restrictions concerning residence, grain supply, and education) caused military enterprises in the interior to operate like vast families, with most of the employees' children working alongside their parents. These 'families' tended to turn in on themselves and thus the quality of the workforce deteriorated rapidly. Moreover, young urban workers were unwilling to transfer to remote military plant, leaving these plants with no option but to recruit locally, including among the offspring of existing workers. In these circumstances, it was inevitable that the quality of the workforce should deteriorate.[19]

The government has taken steps to halt the outflow of workers from military enterprises, chiefly by increasing salaries, arranging opportunities for overseas travel, and offering rewards for the development of outstanding civilian products or technology. However, the results of such initiatives have been patchy.[20]

The fourth impact of the reforms is a fall in technical standards. The stated purpose of the policy of converting to civilian production is to use profits earned from the manufacture of products for the civilian market to subsidize and improve the research, development, and production of military equipment. But lack of funds and the other problems discussed above have made it difficult for the authorities to maintain the production of military equipment.

The authorities have issued orders that essential military production lines must be retained, but military enterprises often find themselves torn between their own survival and the necessity of obeying government orders. Some of them have ignored government requirements and converted military production lines to civilian production, making it impossible for them to fulfil their military production obligations. Other plants have obeyed orders and retained military production lines, but have mismanaged them and deprived them of adequate funding. These kinds of problems, added to labour difficulties and outdated equipment, have cast doubt on the ability of military enterprises to give priority to military production and use profits from civilian production to subsidize the military sector.[21]

IMPLICATIONS OF THE REFORMS

Generally speaking, conversion to civilian production consists of 'political, economic, and technical measures for assuring the orderly transformation of labour, machinery, and other economic resources now being used for military purposes to alternative civilian uses'.[22] The basic assumption underlying this policy is that the technical level of the civilian production ought to be approximately equivalent to that of the original military products. Only in that way will the human and material resources be fully utilized and a drop in technical levels be prevented.

The PRC authorities acknowledge this aspect of the conversion to civilian production, and they stress the need to choose products that fully utilize the technological and manpower advantages of the military plants. They also intervene to assist plants that want to make the conversion, helping them to gain a foothold and ensuring that their choice of products is less random.

But if we judge the programme according to the criteria mentioned above, we are forced to admit that, despite the fact that it has ensured the survival of some military enterprises, the conversion to civilian production has not been successful. It is clear from PRC reports that most of the civilian products manufactured by military enterprises are everyday commodities. According to one source, the types of products may be roughly broken down as follows: electric fans and air conditioners (30 per cent); refrigerators (20 per cent); bedding, clothing and footwear (20 per cent); pharmaceuticals (10 per cent); and machinery and electronics products (20 per cent).[23] Another source claims that only 20 per cent of enterprises have made use of their existing technologies in making the conversion.[24] In other words, most of the civilian products are low-tech, labour intensive items with low added value. They are little different from products produced by ordinary enterprises, resulting in needless competition for markets, raw materials, and capital between the military and the civilian sectors and causing duplication of development.

The PRC authorities are worried that in an effort to ensure their own survival and meet market demands, too many military enterprises are choosing products far below their current technological capacity, causing a deterioration in technology, equipment, and manpower skills. For most enterprises, the conversion to civilian production means making a completely fresh start, discarding their existing technology, allowing their skilled employees to drift away from the interior to the more developed coastal provinces, and selecting new civilian products to manufacture.[25]

This is a sign that the production of high-tech civilian products by military plants has encountered serious obstacles, leaving them with little alternative but to start afresh. One such obstacle is market demand. The reforms have brought about an expansion of the market for consumer products and the prices for these products are now subject to market influence. In a country like the PRC, with a huge population and low average income, the demand for low-tech, labour intensive consumer products is high. When it is a matter of their own survival, military enterprises naturally tend to follow market demand.

Indeed, it would not be easy for military plants to manufacture high-tech civilian products, as the necessary technical renovation would require substantial reinvestment,[26] and lack of funds makes it impossible for the government to seek an all-round solution. Anyway, the market demand for such products is by no means large and it would be difficult for military plants to manufacture them on an economic scale.

The third obstacle to the manufacture of high-tech products concerns technology. As far as the domestic market is concerned, military plants may be technically more advanced than civilian enterprises, but from a global point of view, military enterprises are in no way technologically superior. Moreover, technological progress is continuing apace in the outside world, and products manufactured using comparatively superior technology are still likely to be uncompetitive. This problem is expected to be exacerbated as China's market becomes more open. What is more, the equipment used by military plants is already out of date, and is unlikely to be suitable for manufacturing more sophisticated civilian products.

Most military plants that are manufacturing everyday items are operating according to the rules of the market. Although the authorities may consider their choice of products to be random, the enterprise operators themselves recognize that their most important responsibility is to ensure their factory's survival, and possibly make a profit. So in view of the government's inability to provide any subsidies and their own limited management experience and lack of market information, the enterprises have no choice but to manufacture products for which there is a large current and ongoing market demand.

One salient fact that emerges from the above description is that the PRC's defence industry is far too large. Scholars in the PRC have come to different conclusions regarding the optimum size of the country's defence industry. Some believe that in order to deal with limited wars and possible outbreaks of fighting, and to preserve the industry's current R&D capability, it is only necessary to preserve between one-fifth and one-fourth of current production capacity and one-third to one-half of current research capacity.[27] Other scholars estimate that the PRC only needs approximately 15 per cent of its current military industrial complex to satisfy manufacturing demand.[28] These estimates underline the excessive size of the country's defence industries.

ON THE HORIZON

There are two developments on the horizon which will have a major impact on the PRC's defence industries. These are the promotion of a 'socialist market economy' within China and the PRC's application to join the General Agreement on Tariffs and Trade (GATT).

The Chinese Communist Party (CCP) announced its plans to establish a socialist market economy at the party's Fourteenth National Congress in October 1992. The proposal reflected Deng Xiaoping's call for further deepening of the economic reform effort during his tour of southern China early that year. It stressed that economic planning was not necessarily socialist, as capitalism also had economic plans; neither was the market necessarily a capitalist phenomenon as socialism also permitted the existence of markets. Economic planning and markets were no more than means to adjust the economy. This announcement indicated that as the economic reforms developed, capitalist elements would play an increasingly important role in the PRC economy.

The increasingly capitalist nature of the economy will have an impact on the PRC's military industrial enterprises in two ways.[29] First of all, the military industrial system will inevitably be caught up in the expanding market, and the survival of enterprises will depend on their ability to compete with other firms. Second, there is likely to be an open debate over whether military products are commodities and whether military industrial enterprises should really be independent legal entities.[30] In these circumstances, the scope of economic planning will diminish and military industrial enterprises will have to purchase most of their materials in the market. In the meantime, management methods will have to change.

This twofold impact will also affect the PRC's defence budget and the organization of the military. If military products are acknowledged to be commodities and military industrial enterprises to be independent legal entities, purchasing prices for military products will have to rise considerably so as to enable the enterprises to renew their equipment and to ensure their willingness to undertake military contracts. A rise in the purchasing prices of military products would affect the allocation of the defence budget, as the authorities would have to allow for a considerable increase in funds for purchasing arms and equipment and for supply and maintenance expenses. The government would then either have to increase the defence budget or face more cutbacks in the military, though the latter would put an added burden on local governments and enterprises responsible for allocating employment to the demobilized troops.[31]

Membership of GATT would require China to speed up the opening of its markets, and this would present the country's defence industries with an even greater challenge.[32] In the first place, the industry would have to abolish certain protectionist measures and operate according to international criteria. This would have the effect of destroying the industry's closed, segregated system of organization, but it would also have some negative consequences. Second, comparatively high-tech civilian products are only in the development stage at the moment, and they are not likely to be able to compete with similar products manufactured by more advanced countries. Third, enterprises manufacturing low-tech, labour intensive products are also likely to have problems because they may violate intellectual property rights and trademark regulations. Such enterprises are likely to have to adopt a new strategy after the PRC joins GATT. Finally, GATT entry would affect those enterprises engaged in the service sector. All of these factors will present challenges to the PRC's defence industries in the future.

NOTES

1 For a discussion of the PRC's economic reforms see Edward K. Sah, *Zhonggong shihnian jinggai de lilun yu shijian* (The theory and practice of the Chinese Communists' ten years of reform), Taipei: Institute of International Relations, National Chengchi University, 1991; and Peter Nolan and Dong Fureng (eds), *The Chinese Economy and Its Future: Achievements and Problems of Post-Mao Reform*, Cambridge, Mass.: Polity Press, 1990, pp. 63–71, pp. 139–60.

2 The PRC media are not exact in their use of terminology where defence industries are concerned, and these industries are often referred to as 'military industries' (*jungong*). This term may even include maintenance plants controlled by the PLA General Logistics Department, which do not come under the State Council. See *Dangdai Zhongguo jundui de houqin gongzuo*, Beijing: China Social Science Press, 1990, pp. 592–601.

3 Benjamin C. Ostrov, *Conquering Resource*, Armonk, NY: M. E. Sharpe, 1991, pp. 62–4.

4 These ministries were reorganized again in 1988 and 1993. See, Paul H. Folta, *From Swords to Plowshares? Defense Industry Reform in the PRC*, Boulder, Colo.: Westview, 1992, pp. 202–4.

5 Shi Shiyin, 'An Ideological Weapon in the Reform of Military Industry: A Tentative Understanding of Comrade Deng Xiaoping's Writings on the

Subject', *Junshi jingji yanjiu* (Defence economics studies), Wuhan, 1993, no. 3:57–9.

6 You Qianzhi, Che Xin, Yang Jianrong, *et al.* (eds), *Zhongguo guofang jingji yunxing fenxi* (The workings of China's defence economy), Beijing: China Finance and Economics Press, 1991, p. 106.

7 Ibid., pp. 106–9.

8 Zhu Qinglin, 'Thoughts on the Conversion to Civilian Production in China's Defence Technology Industry', *Junshi jingji yanjiu*, 1992, no. 4:51.

9 For an account of the 'conversion to civilian production' policy, see Folta, *From Swords to Plowshares?*, pp. 52–8.

10 For the transitional stage in the development of civilian products see Sun Zhenhuan, *Zhongguo guofang jingji jianshe* (The development of China's defence economy), Beijing: Military Science Press, 1991, pp. 28–34. The periodization used in this article is that preferred by PRC scholars.

11 Ibid., p. 28.

12 Ibid., pp. 28–9.

13 Jiang Baoqi, Zhang Shengwang, and Ji Bing, 'Some Questions Concerning the Strategic Adjustment and Structural Reform of the Defence Industries', *Jingji yanjiu* (Economic research), Beijing, 1988, no. 12:63; Liang Wenjun, 'Strengthen the Management of State-owned Military Industrial Assets, Promoting the Reform of the Defence Technology Industry', *Junshi jingji yanjiu*, 1992, no. 10:56.

14 Personal interview, July 1993. Li Yintao, 'A Tentative Analysis of the Environment in which the Defence Technology Industry Can Survive and Develop', *Junshi jingji yanjiu*, 1992, no. 1: 46–50.

15 Zhu, 'Thoughts on the Conversion to Civilian Production', pp. 53–4.

16 Yang Zhengguo, 'A Discussion of Certain Problems Concerning the Transformation of Management Mechanisms in Military Industrial Enterprises', *Junshi jingji yanjiu*, 1993, no. 2:54.

17 Song Zhenduo, Ji Bing and Yu Liankun, 'The Plan and the Market in the Conversion to Civilian Production', *Junshi jingji yanjiu*, 1992, no. 5:62.

18 Zhu, 'Thoughts on the Conversion to Civilian Production', p. 53.

19 You *et al.*, *Zhongguo guofang jingji yunxing fenxi*, p. 119.

20 Personal interview, July 1993.

21 See note 15 above.

22 This is the definition given by Seymour Melman and Lloyd J. Dumas, cited in Mel Gurtov, 'Swords into Market Shares: China's Conversion of Military Industry to Civilian Production', *China Quarterly*, no. 134, June 1993, pp. 213–41.

23 Xiao Changjin, 'Some Problems Concerning the Conversion to Civilian Production by Military Enterprises', *Junshi jingji yanjiu*, 1992, no. 5:44.

24 Ibid., p. 45.

25 Regardless of whether making a fresh start is or is not part of government policy, many overseas scholars believe that this is what is happening.

26 Jiang, Zhang, and Ji, 'Some Questions Concerning the Strategic Adjustment', p. 65.

27 Ibid., p. 64.

28 Yang Jianping and Wang Luhua, 'Thoughts on Defence Technology Industries Operating the Strategy of Combining Military and Civilian',

Liaowang zhoukan (Outlook Weekly), Beijing, 1989, no. 48, 27 November, p. 20.

29 Liu Huamian, 'A Tentative Discussion of the Military Economy under the Socialist Market Economy', *Junshi jingji yanjiu*, 1993, no. 1:4–10.

30 For more on this debate, see Yang, 'A Discussion of Certain Problems', pp. 53–6.

31 For the time being, it is impossible for PRC to raise the prices of military products or treat military enterprises as independent legal entities, so the only options open to the authorities are improving management efficiency and encouraging the formation of enterprise conglomerates. See Song Wenge, 'Thoughts on the Development of Military Enterprise Conglomerates', *Junshi jingji yanjiu*, 1992, no. 3:20–21.

32 Fang Qiao, 'Cooperating Sincerely with Friends Amid Peace and Development: An Interview with Tian Ruizhang, Deputy General Manager of the China Northern Industrial Conglomerate', *Zijing Zazhi* (Bauhinia), Hong Kong, August 1993, pp. 41–3.

Part II

The impact of external defence policy

6 China's arms sales

François Godement

Arms sales from China, or arms transfers, since some of these transactions have been donations, are an on-and-off topic because the amounts involved have fluctuated enormously, and also because their political or nuisance value is not always assessed at a very high level. Up to the early 1990s, it has been very much an 'on' topic, because China is the only state that exports missiles of various ranges, and because its conventional arms sales, regardless of their actual quantity, have had a marked impact on the developing world, if not on the armies of industrialized countries. China, moreover, has been consistently ranked among the top five arms exporters to the developing world since the early 1950s – but only as the fifth member of this league until 1980. Up to the mid-1970s, almcst all of it was made up of support to Third World revolutionary movements and new Communist states; most of it was not sold but donated, and the greatest part of the equipment involved consisted of small arms and field communication equipment and other supplies, for which China gained a reputation as a rugged producer. However, their politico-strategic value, or more aptly their nuisance value, was almost limitless in the era of insurgency and people's warfare, and this was sometimes closely linked to China's own participation in Asian land wars such as the Korean and Indochinese conflicts. According to a recent review of the topic, China trained 23 liberation movements, and exported small arms to 45 countries[1] throughout the world, although one must hasten to add that China more often than not also disappointed its revolutionary protégés. Then, the issue was not proliferation or arms trade, but support for movements which antagonized the West, and also the Soviet Union. It is on the strength of arguments (but not proven facts) regarding Chinese arms deliveries by sea to the PKI, Indonesia's Communist party, that a military coup was organized in 1965 which changed the course of Indonesian politics. During the most ebullient years of the Cultural

Revolution (and especially 1968), railroad convoys to Indochina were often hijacked through Southern China by rival Red Guard groups who then used the weapons (Russian and Chinese) for their own agenda: these unforeseen consequences inside China almost split the country, until commanders of the PLA reacted to these trends at Wuhan in July 1967, holding hostage an envoy from the Centre in Beijing, and forcing the demilitarization of Cultural Revolution factions. Although the quantities involved were never insignificant, it was the revolutionary upheaval potential that these sales permitted that mattered, not their technology.

China's arms transfers before 1980 also tended to rise in bursts linked to external conflicts with the United States (in Asia especially) or the Soviet Union (in Africa), as R. Bates Gill has noted,[2] and the only instance where he does not perceive this trend (1958–9) is the era of the Great Leap Forward, when Chairman Mao himself upped the stakes in confronting the United States and started to challenge the Soviet Union. But today, in an interesting twist of post-modernism, Poly-technologies, the PLA's salesman, boasts with some implicit irony of having sold US$200 million worth of semi-automatic guns on the United States market since 1987.[3] The political significance of this is minimal, except perhaps to domestic gun-control groups in America. In fact, given the worldwide availability of many small arms and the cutthroat price competition which the fall of the Soviet Union has provoked, most small arms today would seem to fall more into the category of consumer goods than that of genuine arms transfers. If one was to believe several recent American sources, this would be, in 1992–3, the only problem remaining about Chinese arms exports, since they have seemingly plummeted to a historical low point of US$100 million in 1992.

THE 'MODERN' ERA: 1977–87

In between those two extremes, China has had a first peak period for its weapons sales in the 1980s, thanks to the Gulf War between Iran and Iraq and related sales to the Middle East. This is really the 'modern' period in Chinese arms sales that one expert dates from 1977,[4] but that in any case also predates most known instances of Chinese proliferation of advanced technologies, as opposed to arms transfers involving less advanced technologies: the Iran–Iraq war ended in 1988, and most cases of proliferation date from that time, with some deals reportedly starting in 1987. One of the major problems involved is that figures almost never match between sources, and are in fact impossible to pin objectively in a very elusive area. Thus, Chinese conventional arms sales rose from

US$110 million in 1977 to US$2,553 billion in 1987 according to SIPRI,[5] while other sources, such as the Congressional Research Service and Richard F. Grimmett's statistics, arrive at a figure of US$5 billion for 1987.[6] But nobody denies that 1987 was a peak year for Chinese arms sales and that, overall, China unloaded large quantities of conventional armour (for instance, 1800 T-59 MBTs and 1740 T-69 MBTs), surface to air missiles (SAMs) and artillery to Iran and Iraq during their war. Conversely, CRS sources emphasize the decline in Chinese arms sales after 1987, while SIPRI sources suggest more continuity in these sales for the same period. Should we follow Gerald Segal's assessment of this discrepancy, suggesting that 'the Bush administration's . . . official policy was to suggest China was less of a problem for international security'?[7] This might, by inference, lead one to suppose that the Reagan administration's policy before 1988 was, by contrast, to overemphasize the size of China's strategic threat, since CRS statistics were higher, rather than lower, than SIPRI figures. Or is it just a case of American intelligence sources being more able to track down arms transfers than the private and independent institute that SIPRI is? SIPRI, however, has attempted to test its own statistics against the first returns of the new United Nations Register of Conventional Arms.[8] In so far as Chinese arms exports are concerned, the results are disappointing or reassuring, depending on one's expectations. China – along with Syria – did not participate in the 150–0 UN resolution on Transparency in Armaments in December 1991. It belongs, however, to the group of 8 nations, among the leading 15 importers of arms, that did report transactions to the UN Register in 1992. There were Chinese exports that were declared by China to the UN Arms Register and which had not been identified as such by SIPRI: most notably, the sale of 106 artillery guns to Iran. But there were also, and in larger quantities, sales that had been identified by SIPRI and which were not declared to the UN Arms Register: fighter aircraft (40 F-6, 21 F-7M) sales to Bangladesh as well as sea-to-sea, short-range missiles (Hai Ying-2L) in 1989 and 1992; planes (F-7M, Y-8 transport aircrafts) and helicopters to Myanmar in 1990–92; and also large sales to Pakistan (98 A-5 fighters in 1984, 40 F-7M in 1988, 40 F-7P in 1992). Finally, there were even larger sales to Thailand that were not reported, often by either party, to the UN Register: 9 Dauphin helicopters (or more accurately, Z-9 Chinese models made under French licence) in 1992, 450 T-69 tanks in 1989–92, 4 C-801 launchers in 1991–2, 25 fire control radars also in 1991–2. In other words, China was reasonably forthcoming towards the principle of the UN Register itself, but possibly more evasive when the time came to declare the details of its transactions. The SIPRI data, on average, do seem more complete and comprehensive than other sources, even if there is no guaranty of their complete validity.

Throughout this debate on China's real figures for arms sales, there is of course a degree of hypocrisy. After all, the UN Register, for example, does contain disclosures about sales by advanced countries that were not previously known, even if they were not controversial: for example, some French artillery sales to Saudi Arabia. And China is not the only country that sold weapons to two opposed belligerent states. Recent studies about such a neutral and value-driven country as Sweden have shown that it did not, in fact, live up to its self-proclaimed restrictions on arms sales. Whereas Swedish regulations prevent selling arms to countries involved in armed conflicts, or about to start one,[9] it has sold weapons simultaneously to India and Pakistan in the 1980s, as well as to Israel, Syria, other Arab states, and Iran.[10] This statement is not so much to put Sweden on the same footing as China, but to show that ambiguities abound, and that there seems to be no virtuous actor in this game. In a 1991 retort to US criticism, the then Chinese president, General Yang Shangkun, derisively cited a Chinese proverb: 'Only magistrates are allowed to light fires; ordinary people are not allowed to light lamps'.[11] This was obviously a far cry from the militant rhetorics of the Maoist past: in essence, Yang claimed freedom of trade for China, rather than any specific goal or strategy.

A remarkable aspect of these large conventional arms sales by China in the 1980s is that they have never been appraised as likely to upset any regional balance. With few exceptions – concerning mostly weapons sold at a later date – Chinese arms sales, like their predecessor from the 1950s, were assessed more for their political nuisance value than for their actual potential. The available literature is often full of deprecative comments regarding the 'poor quality' of these weapons, suggesting that the motivations of the buyers are second rate: they are either Third World customers filling up their armies with cheap, low-tech arms which more likely than not they will not use in actual international conflicts (Egypt, Pakistan, Iraq and to a lesser extent Iran, Thailand), or 'Fourth World' states attempting to set up regular armies (Somalia, Sudan, Bangladesh, Zimbabwe) or 'pariah states' which have nowhere else to go: Libya, North Korea, Burma today, as well as the Khmer Rouge.[12] One of the most telling cases is Iran during its war with Iraq. Although it came close to depending on China, which supplied 22 per cent of its armaments between 1982 and 1989, it has more often than not refrained from engaging them against Iraq, using Soviet and other supplies for this. Similarly, Iraq would later leave unused its large number of Chinese tanks and MBTs against the Allied coalition, and with good reason. In these instances, arms purchases from China appear to bolster inventories, to shore up the leaders' claims, and possibly belief, in the magnitude of their armies, and also to create an impression with worldwide public opinion: that impression, however, is a bluff that can easily backfire. In 1990–91, media claims that Iraq possessed one of the world's most potent armies rested in the main on these large, but inferior inventories, and thus served ultimately to mobilize support against Saddam Hussein.

Thailand is also a country that has come to supplement its Western arms purchases with a degree of Chinese weapons, in what appears mostly as a balancing act to maintain its independence, to appease China and possibly to supply neighbouring forces such as the Khmer Rouge: but it is also the source of much negative comment about the inferior quality of these Chinese weapons.[13] On the Chinese side, these sales may also have had a geopolitical intent, if it is true that Thailand obtained many weapons for rock-bottom prices.[14]

An exception to this rather low assessment of the effect of Chinese weapons sold between 1977 and 1987 may be the Silkworm's various versions sold in the Middle East: both the shore-to-ship and after 1990 the Haiying-2 ship-to-ship versions. Although not technically prevented by any restrictions, these sales were treated by China itself as borderline cases of proliferation, since it denied them and later used North Korea as an intermediary during the Iran–Iraq conflict. Again, Silkworm missiles were more useful for their political and economic nuisance value – scaring cargo trade away from the Gulf – than for their military value per se. In September 1987, several launches of Silkworms by Iran in the Persian Gulf brought havoc to traffic, and forced the United States to patrol the area: three years later, some Silkworms were fired by Iraq against the USS Missouri during the second Gulf War; one was intercepted by the HMS Gloucester – the one and only anti-ship missile (AShM) ever brought down in combat since these weapons have been invented.[15] But Silkworm missiles had, in fact, been sold to other countries in the region: Pakistan from 1981, Egypt from 1984, as they were also sold at the same time in large quantities to Bangladesh. The pattern suggested mostly an arms-for-cash strategy, with realistic misgivings on China's part about Iran's possible use of these missiles, rather than any far-ranging goal.

If the assessment of Chinese arms sales was to end with 1987, one would say that China had engineered, in the space of a decade, a spectacular turn from mostly ideologically, and sometimes strategically motivated, transfers to commercially designed sales. In the same period,

Africa had declined as a market, the Near and Middle East had taken on tremendous importance, and Asian customers beyond the traditional Communist states and movements had begun to appear. None of it, however, seemed to suggest a particularly strong motivation, other than financial gain. It is nonetheless true that China also practiced a degree of selectivity in its sales where its own strategic interests were at stake. China supplied both the Nicaragua Contras and the Afghan resistance with SAMs; in Africa, it sold more indiscriminately artillery and naval crafts, but not missiles or planes; in Asia, the list of customers was much more restrictive, with Thailand as the only ASEAN nation represented. But was this completely China's choice?

Studies detail the defects of China's arms industry – its shortage of young scientists, its neglect of long-term research and development (R&D), its preference for over-the-counter package acquisition of foreign technologies (electronics, avionics), as well as the commercial competition between Norinco and Polytechnologies, the two leading salesmen belonging respectively to the defence industry and to the PLA itself.[16] Whereas the Norinco logo (*Beifang gongsi*), invented first as a fictitious unit for the Fifth ministry of Machine-building, is a symbolic tribute to the former Northern Army of pre-1911 fame, the Chinese characters for Poly (*Baoli*) were synonymous with the PLA's urgent desire to 'retain profits' rather than see the defence industry monopolize them in the export field.[17]

THE TURN TO PROLIFERATION

According to SIPRI data, Chinese arms sales between 1987 and 1989 declined by two-thirds, falling under a billion dollars for that year.[18] Its relative share of the developing world arms market (sales to industrialized countries remained insignificant) declined as well, from 8.8 per cent to 7 per cent. Was this what one expert in the field called 'the death of an (arms) salesman'?[19] The Iran–Iraq ceasefire in 1988, the lessened conflicts in the Third World, thanks to a new Soviet Union that had already changed its foreign policy under Gorbachev but was still in control of its satellite countries, the rise of new arms producers, all led to a decline of the armaments market. Furthermore, commercial pressure from the arms industries in advanced countries, where defence expenditures were peaking or starting to decline, also made it possible for developing countries to seek more expensive armaments: coupled with the new prosperity in Asia, this meant that the only region where military expenditures kept rising was East Asia – but the proceeds went to Western manufacturers or new local producers rather than to China.

Pakistan, a staunch customer of China's warplanes, is now ready to buy fighters from Central Asian states, which will obviously be of Russian origin.[20]

These trends were countered by China in two ways: by offering newer and more sophisticated weapons, and by resorting to more controversial sales, especially of ballistic missiles and/or technology. China is of course still developing new arms systems, even if cooperation from all Western partners is not as forthcoming. It is – once again – trying to fill the gap in its air force by developing its own advanced jet fighters by the year 2000, according to Zhu Yuli, president of Aviation Industries of China.[21] According to the same industry figure, China now has an 8-year programme to modernize its aviation industry as a whole, with an $800 million target for export earnings in 1995.[22] It has entered the preliminary stages for building two Kiev-class aircraft carriers by the year 2005, earmarking an initial budget of 10 billion yuan for this purpose, and is already training pilots for this.[23] Indeed, it has acquired inflight refuelling capacity for its new naval J-8 II fighter from Iran, and is already marketing this feature for export customers. From designs by the late Gerald Bull, of Iraq fame, and with parts adapted from a disused US artillery system, Norinco has apparently put in production the world's longest range artillery gun, a 203 mm piece with a 50 km radius, which is clearly designed for export.[24] Most of these developments are rather distant, however, and do little to bridge the gap, in the short term, with competitors for arms exports, prompting the judgement by some that the distance between Chinese and Western or Russian technology is increasing, not decreasing. A question for the future lies with China's capacity to transfer technology from Russia. The recent and publicized indication that, due to budgetary constraints, the PLA might buy just a limited number of SU-35 fighters could mean that China is largely interested in eventual offers of technology transfer, or in attempting yet another effort of reverse engineering. On a commercial level, one fails to see the Russian advantage in allowing a competitor to catch up with their own products. But in strategic terms, there are those in Moscow, and in the military-industrial establishment, who would like to 'look East' as a complement or a replacement of Western influence: this might bring in the future a new source of weapon technology to China.

Two developments have already been successful for export, however: the F-7M Airguard fighter and the C-801 successor to the Silkworm missile. The F-7M fighter has been exported to Bangladesh, Iran, Myanmar, Pakistan (China's biggest customer for this model as well as for its predecesors), Sri Lanka, and Zimbabwe. The C-801 has been sold to Bangladesh and Thailand. Both of these weapons systems benefited, willingly or not, from Western technology: the F-7M through improved British avionics and other equipment, the C-801 because it appears very similar in design to the Exocet AShM. In fact, new versions of China's fighters and interceptors were being designed at the same time with the cooperation of US aerospace firms, partly under the 'Peace Pearl' programme: other, smaller projects, involved Aeritalia or Thomson. These projects were scrapped after 1989 because of the aftermath of the Tiananmen crisis, and the ban on military contacts with China that ensued. Similarly, China's Jiangdong class frigates (sold to Thailand in 1990), which succeed the Jianghu class, are equipped with German turbines, a fact that has often been underlined with regret by Taiwan military sources. New Chinese destroyers are also developed with GE turbines. All these cooperative actions, however, would imply either that the Western governments authorizing them did not think that these weapons were potentially destabilizing, or that they chose to boost profits first, and ponder strategic consequences later.

In at least one case of alleged cooperation, however, one cannot help but wonder. Reports about Israel–China cooperation have surfaced since 1988,[25] placing the start of this cooperation as early as 1981. Shells and artillery, including the 105 mm gun intended for the Merkava tank, contacts through the Shaul Eisenberg offices in Beijing, probably date from the early period of this relationship: when Israel was conducting a 'quiet diplomacy' to woo China.[26] Whether or not Israel cooperation went as far as to help designing the guidance system of the CSS-2 intermediate range ballistic missiles (IRBMs), the general results would appear to have backfired by 1987, when China sold these CSS-2 to Saudi Arabia, with the entire Middle East in their range. Could it be that Israel's cooperation with China, which seems to have grown in the late 1980s, was motivated as much by the desire to contain and prevent such developments in the Middle East, as by the quest for diplomatic normalization and funding its own arms industry? In that case, the CSS-2 sale to Saudi Arabia appears to have been a 'multiple hit' in . . . multiple ways.[27] Not only did it bring a handsome profit to China's Polytechnology group: 'bu shao' (not little), as Deng Xiaoping is said to have remarked in approving the deal,[28] and cadres in the rocket industry would be overheard saying 'gande piaoliang' (beautifully done), according to another account.[29] Those were the years when the 'Dengist companies' were doing their best to bring in the cash for the Four Modernizations programme. But it also helped China secure diplomatic recognition by Saudi Arabia, formerly a key partner for Taiwan, in July 1990. And perhaps, it helped draw the Israeli military further in their wish to control the China factor in the Middle East. Several times, Israeli officials have boasted of having been able to stop China's missile sales to the Middle East thanks to their own good Beijing connection.[30] Others, however, have judged that China got the better part of the secret deal: 'The secrecy most likely worked to Israel's disadvantage, as Israel was unable to benefit from the relationship politically, its primary aim in pursuing the relationship in the first place'.[31] Meanwhile, and especially after the CSS-2 sale, and even more after Tiananmen 1989, Israel went further in, drawing the anger of US authorities and several CIA reports that it had sold up to $10 billion of weapons and military technology to China,[32] culminating in the Patriot controversy. One might find a minor corroboration of these trends by looking at Israel's early 1994 effort to woo North Korea: certainly, North Korea's diplomatic recognition of Israel has no value whatsoever on a world scale, but perhaps a stop to North Korean arms supplies to the Middle East would justify Israeli concessions? If North Korea can get Israeli envoys to Pyongyang, one imagines that the Chinese have gained an even bigger leverage, thanks to their proliferation deals in the Middle East.

On balance, it would seem that the two states had been working at cross-purposes for some time. China had been designing ballistic missiles mainly for export (and code-named M, vs. DF for PLA domestic use) since 1984, when domestic defence appropriations were severely cut. The M-9 programme, for instance, was started by the First Academy in April 1984: it became the contractor for all of China's 'tactical' missiles, e.g. with a range under 1,000 km (as opposed to the 300 km ceiling for tactical missiles under Western and 1987 Missile Technology Control Regime (MTCR) criteria). The first missiles were openly displayed for sale at the 1986 Asiandex show in Beijing; these same M-9 missiles would be sold to Syria, then Israel's mortal enemy, and Syria would allegedly finance some of China's missiles projects, while also buying North Korean Scuds which incorporate Chinese technology.[33] Meanwhile, after the US placed sanctions on China for its missile transfer to Pakistan, then lifted these sanctions, China in return pledged to abide by MTCR restrictions in February 1992, although still not participating in the regime discussions themselves. China promised several times to refrain from further missiles sales to the Middle East. It had also received Mr Arens, the Israeli Defence Minister, for secret talks, and finally established diplomatic relations in 1992. Israeli officials claim that China has promised not to sell missiles to Syria 'in the future',[34] and also say there is no proof of delivery of any M-9 missile to Syria, while American sources characterize the deal as 'on hold'.[35]

Meanwhile, China also embarked on a course of ballistic missile proliferation with Pakistan, in what was at least as much a strategic venture as one for cash. While border negotiations with India falter in spite of formal normalization of relations, China is a major arms supplier to every country of South Asia except India: Burma, Bangladesh, Nepal, Sri Lanka, and of course Pakistan. China has sold M-11 (CSS-7) missiles since 1991, and rather than denying the sale claimed that the 'tiny amount' sold had a tactical range falling just under the ceiling set by the MTCR, according to the Chinese envoy to the United States. After counter arguments by the United States, Pakistan denied those sales, as well as cooperation between the two countries on missile production which has helped Pakistan bolster its indigenous capacity. But whether as Pakistani Chief of Staff or as a retired but influential figure, General Mirza Aslam Beg has repeatedly blown the whistle on Chinese–Pakistani cooperation, boasting repeatedly about the delivery of the M-11 missiles.[36] Just as in the case of alleged nuclear cooperation between China and Pakistan several years ago, these missiles sales have brought Washington to trade sanctions against China, before lifting or ending them. China did give up its plan to sell a Han class nuclear submarine to Pakistan, which would have served to balance the Indian submarine fleet. Tension rose again in 1993, when the United States mulled trade sanctions in the spring,[37] then decided on a two-year ban for exports of sensitive US technology to China, officially estimated to cost the United States itself between US$400 and US$500 million in lost business for each year.[38] One result of these sanctions has been the announcement by China that it had 'to reconsider its commitment to the MTCR'.[39]

At the same time, China also provided nuclear research reactors, albeit of a small size, to three countries whose names look like a list of candidates for the Islamic bomb: Pakistan in early 1990, Iran in mid-1990, and Algeria.[40] Again, these efforts stemmed from earlier decisions, however: the agreement to sell Algeria a research reactor had apparently been discussed as early as 1983. The construction itself was started in 1987–8, and became known to some – but not all – US authorities in 1988. The US made the news public and forced a halt to this programme in 1991, during the last stretch before China ratified the NPT.[41] China finally announced its decision to ratify the NPT in late 1991, and acceded to the treaty in March 1992. These previous reactor sales, although clearly of a proliferatory nature, could not be characterized as legal violations, since they were performed before that date.

But whether the motive for the sales was business or otherwise, the rationale for the purchases on the other side could be only familiarity with nuclear technology as a whole. And the heavy water process involved created the possibility of producing plutonium.[42] In 1992, China signed agreements to sell energy-producing reactors to Pakistan, Iran, Bangladesh and Iran. Overall, the volume of China's exports in nuclear industry increased by 65 per cent between 1985 and 1990,[43] and presumably has kept growing since. They are of course subject to International Atomic Energy Association (IAEA) rules: a recent study of these problems notes that 'China took some time to learn safeguards in nuclear exports. Its conformist behavior resembles French experience and attitude before 1979'.[44]

There have been other questions recently, notably regarding the North Korean armament drive. In the process of North Korea confronting the IAEA and UN Security Council demands regarding verification of nuclear installation, the Pentagon has revealed that North Korea was working on a two-stage missile programme known as Taepo Dong 1 and Taepo Dong 2. The first stage of the Taepo Dong 2 missile is very similar to the Chinese Dongfeng 2 rocket, prompting suggestions of secret cooperation of China in the North Korean programme.[45]

On balance, these developments need to be carefully assessed. China's ballistic programmes for export were launched before the MTCR restrictions were conceived (and in fact, these restrictions may well have been created with the aim of hampering the existing Chinese programmes). Although these exports took off in 1987, right when the end of the Iran–Iraq war (and therefore of the good years in conventional arms trade) was in sight, they had been planned much earlier: nuclear exports from 1983 (cf. the Algerian example), ballistic missiles from 1984. The most blatant act of regional proliferation was hatched relatively early, too: it was the Saudi CSS-2 deal. The Syrian M-9 deal would seem to have served mostly as a very expensive bargaining chip; the M-11 Pakistani deal involves a wider and longer term strategic cooperation with this country, now China's most important military ally. It is worrying because of other aspects of Pakistan's drive for military parity with India, but it is a moot point whether the M-11 deal actually infringes the MTCR provisions, given the debate about the range of these missiles. China's signing up to IAEA and MTCR rules cannot mean that it will not try and interpret those rules to its own advantage, if it can get away with it. But there is no sign that it is stepping up a large-scale rogue programme of balance upsetting exports. In the Middle East, although it has maintained exports to Iran, it avoids challenging the West directly: it has accepted, for example, the inspection of the Yinhe cargo in August 1993, all the better to require an apology from the US after this inspection proved useless. Although it has sold a lot of SAMs to Iran, it has refrained from installing the C-801 successor to the Haiying-2 sea-to-sea missile, which would have been an obvious candidate for export. This is all the more remarkable as its exports to Iran, on the other hand, have suffered from tremendous competition by the Soviet Union and its CIS successor.[46] Russian exports to Iran moved from zero to US$4.3 billion between 1989 and 1992, while Chinese sales fell from US$3.6 billion in 1985–8 to US$1.1 billion in 1989–92. And China does not seemed to have touched the Iraq market, by published accounts at least, since 1990.

Indeed, the two marked features of China's arms sales in most recent years seem to be, first, the notable lack of large deals and second, the concentration of early 1990s deals in South Asia, rather than in the Middle East. China's arms sales to Burma started reportedly in August 1990, in the wake of what one might call the 'twin repressions' of August 1988 (Rangoon) and May–June 1989 (Beijing), which led to increased friendship between the two regimes. There has been much variation in the figures for Chinese arms sales to Burma, and also speculation about Burma's capacity to pay without resorting to benefits from the drug trade. The first agreement, in 1989, was valued at US$1.4 or US$1.2 billion,[47] although SIPRI adopts a much more conservative figure of 407 million for the period 1988–92.[48] It would seem particularly difficult to put a monetary value for arms sales between China and Burma, given the secretive nature of both defence establishments and the obvious opportunities for many forms of compensation other than cash. China's latest colonial adventure may be here, with many motivations to pick from: opening a land route to the Gulf of Bengal, countering India, profiting from the opportunities of the Golden Triangle, extending its influence from the edge of the Indochinese peninsula.[49] In fact, it is not so much the arms sales to Burma which have raised eyebrows (although the EU has, for example, banned in 1991 arms sales by its members to the Burmese junta), but some strategic side aspects of the deal: China has gained observation posts on several islands in the Sea of Andaman, very close to Sumatra, on the edge of one of the world's busiest maritime route. This, along with the increased arms sales to Bangladesh and other South Asian partners, suggests that General Yang Shangkun's Southern Asia strategy of the early 1990s is bearing fruit. Does this, however, represent world-class influence through the sale of arms?

Throughout the preceding review of evidence, we have shown some scepticism about the strategic decisions and overall threats implied by China's arm sales, or even acts of proliferation. After the revolutionary era and the temptation to foster movements against the two 'superpowers', came the era of arms for cash: the Middle East was an opportune ground, which China was not the only supplier to cover. Lately, the quest for cash seems to have engulfed the Chinese military so much that they have been, alone in the world, more than ready to convert to civilian industries, a welcome consequence of their appetite: it is the 'commercialization of the PLA'.[50] There remain elements of a regional strategy in China's South Asia arms drive, and presumably China would have liked to achieve similar results in Southeast Asia: it is outclassed, however, by many other sellers. Nuclear and ballistic proliferation were launched in an era when China was not held accountable to international restrictions. China then locked horns with the United States on many occasions, often getting away with its initial sales (Saudi Arabia, Pakistan, perhaps even Syria), but having to reduce them to one-shot deals. Just as the United States, and other willing parties, have established a linkage between China's compliance with restrictions in nuclear and ballistic trade and some advantages granted to its economy, China has linked its compliance (or its promise to comply, in the case of the M-11) to the continuation of these advantages, and to a general lack of sanctions against any aspect of its policy. It has therefore adopted a policy of what one might call conditional, or reversible, compliance. A very good indication of this trend lies in China's attitude to the Permanent Five meetings regarding conventional arms transfers and weapons of mass destruction, which have been held since July 1991. China decided to stop attending this group in protest against F-16 sales to Taiwan in September 1992: which seems to have stalled any further meetings, although China finally gave a 'semi-positive' response to participate again in the process in July 1993.[51] An embarrassing question for the future must therefore be asked: has China decided, in 1991–3, to join, or follow, most of the existing regulations and guidelines regarding arms transfers and proliferation because it has matured from the arms frenzy of the 1980s, or because US policy towards China, in spite of all the criticism it often receives, has achieved that result? And if international pressure was to slacken, wouldn't China's behaviour also begin to alter?

In the short term, however, prospects are more reassuring. Especially for medium to small powers, arms sales traditionally coincide with a variety of motivations, rather than with a grand strategy. China seems to be no exception. We should not focus on the cash-or-strategy debate, which seems pointless. Ever since 1977, China has been pursuing a mixture of both. It has avoided selling to any country that did not have at least a moderately good relationship with China, and although it has resumed many other forms of trade with former enemies such as Vietnam, it is clearly cautious about the reverse effects of its own arms sales. It has kept some customers even though relations were becoming much less important, as was the case with Egypt. It has avoided selling weapons systems that might be used directly in a military challenge to the West (e.g. C-801 to Iran): or rather, it first tried to compromise and hide those sales, and then it renounced them. The Saudi CSS-2 sale appears as a highly successful, but also isolated, venture: and it took place with a regime that can only be described as wed to the West for all practical purposes, so that China may not have had the feeling of upsetting the deck of cards. China seems to play at its best in supplying plentiful, but second-grade weaponry for regional conflicts (Iran and Iraq in 1982–7), and only slightly improved versions of these to poor states it would like to see as clients more generally (Burma, Bangladesh, possibly Sri Lanka): Thailand seems to lie just a little out of range for this prospect, which might have been realistic if the Thai economy did not grow so fast. Mostly, the arms for cash strategy seems to take precedence over domestic weapon development and deployment; conversely, China is never averse to learning from some of its export deals, and doing bits of imitation of reverse engineering to upgrade its own capacities: Egypt, Syria, Iran, and Pakistan would seem to have provided, at one time or another, opportunities for this.

At the same time, some of China's mostly opportunistic endeavours may, one day, solidify into a regional strategy. Gone are the days when China looked to Africa to counter big power influence with aid and weapons. Gone are the days when it was the world's readiest supplier to the Middle East, since Russia is only too eager to perform sure deals in the future. Ballistic missile proliferation seems to have happened rather than to have been planned as a political strategy, whether one looks at the *gaogan zidu* (offspring of high leaders . . .) or at a rather uninhibited marketing drive as explanations. The most discernible trends of recent years point to a narrowing of China's arms for export drive to Southeast and South Asian neighbours, which may be the only avenue presently open to China's arms salesmen. This happens to coincide with China's South Sea strategy: a variety of leaps with countries that are (except Pakistan, of course) largely inconsequential by themselves, but which add up to a chain of Chinese military influence on the southern slope of the Asian landmass.

NOTES

1 According to R. Bates Gill's useful summary of his earlier work in 'The challenges of Chinese arms proliferation: US policy for the 1990s', 31 August 1993, Strategic Studies Institute, US Army War College.
2 Ibid., p. 19.
3 See Tai Ming Cheung's interview with Xie Datong, executive Director of Poly Group Corp, in *Far Eastern Economic Review*, 14 October 1993, p. 68.
4 Richard A Bitzinger, 'Arms to go: Chinese arms sales to the Third World, in *International Security*, Fall 1992, p. 86.
5 *SIPRI Yearbook 1991*, Table 7A2.
6 Cited, for instance, by Richard A Bitzinger in the above-mentioned study.
7 See Gerald Segal, 'China arms transfer policies and practices', paper for the Carnegie Endowment Conference, November 1992.
8 See E.J. Laurance, S.T. Wezeman and H. Wulf, *Arms Watch: SIPRI report on the first year of the UN register of conventional arms*, Oxford University Press, 1993, for the following discussion.
9 See, for example, Sven Hirdman, *Sweden's Policy on Arms Exports*, Utrikesdepartment, Stockholm, 1989.
10 Bjorn Hagelin, *Neutrality and Foreign Military Sales, Military Production and Sales Restrictions in Austria, Finland, Sweden and Switzerland*, Boulder, Westview Press, 1990.
11 Cited by R. Bates Gill, ibid., p. 13.
12 For this categorization, although not necessarily for the same list of countries, see especially Richard A. Bitzinger, 'Chinese Arms Production and Sales to the Third World', RAND note, 1991, pp. 14–17.
13 See Rodney Tasker, 'Order arms: new drill after China's hardware disappoints', *Far Eastern Economic Review*, 4 October 1990.
14 See, for example, 'Thais expected to buy inexpensive Chinese jets', *International Herald Tribune*, 2 May 1989.
15 See 'The coastal missile threat in the Middle East', *Jane's Intelligence Review*, January 1994, p. 26.
16 On the structure of the arms trade itself, see the classic article by John W. Lewis, Hua Di and Xue Litai, 'Beijing's Defense establishment: solving the arms-export enigma', *International Security*, Spring 1991; also Yan Kong, 'China's arms trade bureaucracy', *Jane's Intelligence Review*, February 1994.
17 See Yan Kong, ibid., p. 81.
18 *SIPRI Yearbook 1993*, p. 444.
19 Richard A. Bitzinger, 'Arms to go . . .', p. 93.
20 See 'Decision to purchase aircraft from Central Asian states', editorial in *Jang*, Lahore, 17 May 1994, in BBC, SWB FE/2003/A1, 21 May 1994.
21 See *Asian Defence Journal*, March 1994, p. 93.
22 See *Asian Defence Journal*, August 1993, p. 90.
23 See 'Carriers key to Chinese air power', *Jane's Defence Weekly*, 25 September 1994, p. 23.
24 See 'China's artillery extends its reach', *Jane's Defence Weekly*, 8 May 1993, p. 22.
25 They appeared first in the *Sunday Times*, 3 April 1988.

26 See *Jerusalem Post*, 15 March 1992.
27 See Yitzhak Shichor, *A Multiple Hit: China's missile sales to Saudi Arabia*, Sun Yat-sen Center for Policy Studies paper, Kaohsiung, 1991.
28 See John W. Lewis *et al.*, 'Beijing's . . .', p. 96 for this classic anecdote.
29 See John W. Lewis and Hua Di, 'China's ballistic missile programs', *International Security*, Fall 1992, p. 38.
30 See *Jerusalem Post*, 15 March 1992, and *Middle East International*, 22 October 1993.
31 See Han Xiaoxing, 'Sino–Israeli relations', *Journal of Palestine Studies*, XXII, no. 2, Winter 1993, p. 72.
32 See for example 'Israel sold weapons to China', *International Herald Tribune*, 13 October 1993.
33 See 'Syria, China cooperating on missile development', *Jerusalem Post*, 7 May 1993.
34 See Shimon Peres's account of his talk with Chinese foreign minister Qian Qichen in *Jerusalem Post*, 23 May 1993.
35 See R. Bates Gill, 'The challenge of proliferation . . .', Appendix A, p. 68.
36 He was quoted most recently on this by Ahmed Rashed, 'Bare all and be damned', in the *Far Eastern Economic Review*, 5 May 1994.
37 See *Asian Defence Journal*, June 1993, p. 89.
38 See *Jane's Defence Weekly*, 4 September 1993.
39 See Li Daoyu (PRC ambassador to USA), 'Foreign policy and arms control: the view from China', *Arms Control Today*, December 1993, p.11.
40 See Yan Kong, 'China's nuclear bureaucracy', *Jane's Intelligence Review*, July 1993, p. 323.
41 See 'China A-aid to Algeria: U.S. knew, but didn't act', *International Herald Tribune*, 17 November 1991.
42 See 'Les experts s'interrogent toujours sur la veritable nature du réacteur nucléaire d'Aôn-Oussera au sud d'Alger', *Le Monde*, 9 May 1991.
43 See Yan Kong, 'China's nuclear bureaucracy . . ., p. 324.
44 See Weixing Hu, 'China's nuclear export controls: policy and regulations', *The Nonproliferation Review*, Winter 1994, p. 5.
45 'North Korea grasps at the stage beyond Nodong 1', *Jane's Defence Weekly*, 19 March 1994, p. 18.
46 See interview of Richard Grimmett, *Far Eastern Economic Review*, 26 August 1993, p. 11.
47 See Bertil Lintner, 'SLORC salvation', *Far Eastern Economic Review*, 2 October 1991, and in *Jane's Defence Weekly*, 27 November 1993, p. 11.
48 See *SIPRI Yearbook 1993*, p. 481.
49 See Bertil Lintner, 'The volatile Yunnan frontier', in *Jane's Intelligence Review*, February 1994, pp. 84–7.
50 David Shambaugh, quoted in the *International Herald Tribune*, 'Chinese army gets down to business', 25 May 1994.
51 See *Arms Control Reporter*, 1–94, 707 A-3.

7 Defence modernization and Sino–American relations

Michael Yahuda

The role that the United States has played in the process of China's economic reforms and the modernization of its defences can only be understood within the context of Sino–American relations as a whole. The United States looms very large in most aspects of China's foreign relations and especially on issues to do with strategy, defence and modernization. Indeed these issues touch the core of Sino–American relations as they involve security and economic concerns that are fundamental to the interests of both sides. As such they necessarily reflect the deeper historical and cultural forces that have traditionally shaped relations between these two states that are the product of totally different historical experiences. Although the oscillations in Sino–American relations since World War II may be seen as the product of changes in the international strategic environment, the ways in which each side conducted the relationship reflected these deep differences in the political cultures between them.

This was more evident perhaps as the two states and societies became more closely engaged after Kissinger and Nixon opened relations in 1971–2. In his authoritative account of the two decades since then, Harry Harding has charted a cyclical pattern by which periods of warm and even euphoric embrace have been followed by periods of disenchanted confrontation that in turn have led to reconciliation and then to a renewal of the cycle.[1] Although each side has recognized that deep differences exist between their two societies in terms of values, political systems and levels of economic development, they have not found ways to limit the extent to which these can threaten to undermine interests which they share in common. On the contrary, the Chinese side has tended to accentuate the problem by continually depicting the United States as an imperialist power that is opposed to China and its communist system, while simultaneously demanding the transfer of advanced military technology, the provision of vast economic assistance and unhindered access to the American market as proof of American sincerity in treating China as a friendly and equal power. The Americans in turn have contributed to these problems through a continual redemptionist trend in the way they have tended to treat China's economic reforms as a movement towards capitalism and liberal democracy and in a tendency to caricature contemporary China in either excessively bright or dark colours.

Yet despite these immense problems which give rise to a litany of issues on which Chinese and American governments regularly clash, many aspects of Sino–American relations have continued to flourish and deepen even in the period since the Tiananmen crisis of 1989. Economic and cultural relations have grown in volume, discussions and agreements continue to be reached even on contentious issues. As Steven I. Levine has suggested, all the discordant noises reverberating in Sino–American relations 'are not the rumblings of impending collapse but the sounds of animated engagement between two countries whose relations remain as substantial as they are often conflictive'.[2]

This timely if perhaps optimistic reminder of the enduring substantive content of Sino–American relations should not obscure the fact that some aspects of the interactions have been adversely affected by the 1989 crisis. None more perhaps than relations between the Chinese and American military and American transfers of military-related technology which were abruptly terminated after 4 June 1989 and were only gingerly restarted in late 1993. Even after President Clinton retreated from his previous position and conceded on the most favoured nation (MFN) issue in late May 1994 and separated trade from human rights questions in the conduct of relations with China he still kept on the residual sanctions of forbidding arms sales and the transfer of military technology to Beijing.

These difficulties in the conduct of relations with China reflect the uncertainties in understanding the character of the post-Cold-War era as a whole. It is not as if China alone has been singled out for the pursuit of contradictory and inconsistent policies. But what is significant in Sino–American relations is that the removal of the strategic structures of the Cold War era has brought to the fore the underlying deep cultural and political divides between the two sides. The removal of the strategic overlay that set a framework of priorities which enabled fundamental differences in other matters to be overlooked, or set aside, left these exposed in ways that the systemic differences between the two polities only served to highlight. The ending of the Cold War era also accentuated the ideological divide between Beijing and Washington as the former became increasingly apprehensive about the survival of

Communist Party rule in China and the latter became more confident about its mission to promote the enlargement of the scope of democratization in Asia as elsewhere. The growing significance of domestic factors in the making of foreign policy that is characteristic of the new era has also contributed to accentuating the problems in Sino–American relations. In Washington the increased importance of Congress, lobbyists, single issue groups, and inter-agency rivalries have added to the difficulties in seeking to coordinate policy towards China, and to establishing a coherent institutional structure for administering policy. The importance of domestic factors is also more evident in the conduct of Chinese foreign relations. But here too this has been made more complex as central decision-making has become more fragmented and economic authority decentralized.[3]

This chapter will first present a brief historical overview of the evolution of the American factor in China's defence-orientated economic reforms during the Cold War era. It will examine how the prevailing broader strategic considerations enabled each side better to manage their differing interests despite their divergent cultures and historical experiences. It will then explore the impact of these divergences on the character of Sino–American interactions in the new period and their implications for defence-related issues in China.

HISTORICAL OVERVIEW[4]

For the purposes of analysing the pattern of economic and defence-related exchanges, it is possible to divide the more than two decades of Sino–American relations from 1972 into roughly three periods: first, that of strategic cooperation 1972–82; second, that of commercial interaction 1983–9; third, that of estrangement from 1989 onwards. Paradoxically, economic and military interactions did not fully develop until the strategic imperative for Sino–American relations began to ebb in the early 1980s. In other words it was only during the Reagan Administration that relations began to take off. Indeed it was not until the 'pro-Chinese' Alexander Haig was replaced by George Shultz as Secretary of State, who proceeded to downplay the strategic significance of China and treated Japan as America's most important interlocutor in Asia, that Sino–American relations enjoyed what in retrospect seems their halcyon years. The primary reason was not necessarily that China's strategic role had declined, but that it had been clarified from the American point of view. As indicated earlier, the period of estrangement following the crisis of Tiananmen witnessed a four-year cut off in military relations, but in many other respects the American contribution to China's reforming efforts continued to expand. In sum the American role has been highly complex, but the evolving global strategic context provides a convenient backcloth against which to understand the dynamics of the relationship.

Strategic cooperation, 1972–82

Although the United States and China nominally shared a common strategic adversary in the Soviet Union, in fact they differed in their approaches towards it. For the bulk of this period the Chinese side sought to persuade the USA jointly to confront the Soviet Union, while the USA sought to use the new relationship with China to urge restraint upon Moscow to facilitate the development of détente. Beijing feared lest Washington stand on its shoulders to reach Moscow only to abandon it once the goal had been reached. Washington in turn was concerned lest too close an association with Beijing would exacerbate Moscow's fears of an alliance against it which would undermine its interest in détente and lead to a Soviet–American confrontation. Some American officials favoured selling arms to the Chinese, in part to prevent a possible erosion of Washington's relations with Beijing and in part as a punishment to Moscow for behaviour deemed to be aggressive. But in fact little of significance was sold during this period. Not even the normalization of Sino–American relations announced in December 1978 opened the door to direct arms sales.

Normalization was significant, however, as it allowed the rapid development not only of civilian ties, including the commercial, cultural, and academic, but also facilitated the deepening of strategic links. The key factor in the latter, however, was the Soviet invasion of Afghanistan in December 1979. That paved the way for instituting exchanges and links between the military establishments of China and America and gave the Chinese access to modern military knowhow. Two stations for monitoring Soviet nuclear and missile tests were located in China and the US government agreed to allow the transfer to China of advanced military technology of a non lethal kind. In particular these included a range of equipment to receive data from Landsat satellites, various transport and communications systems, and other technology that was available to any friendly country which would not contribute to the development of weapons of mass destruction. For its part China claimed to be interested in purchasing weapons, but in practice it was constrained by limited finance and indeed by the formal downgrading of the military to last in the Chinese list of the Four Modernizations. Perhaps most important was the decision by Carter in

1980 to extend most favoured nation (MFN) treatment to China coupled with the Chinese reversal of its earlier stand against foreign indebtedness. Further agreements were made to facilitate the more extended range of interactions. Trade increased greatly, but contained little of military value and in the absence of appropriate Chinese laws investment was still minuscule.

Only after the Soviet invasion of Afghanistan did effective strategic cooperation between the two sides begin to take place. But ironically, within two years the strategic basis of the relationship began to unravel. The Chinese calculated that the momentum of Soviet expansionism had come to an end and that the Soviet Union no longer constituted the same immediate threat. The Americans under the newly elected President Reagan had initiated a new military build up. In the circumstances the Chinese once again tested the character of Sino–American relations by putting pressure on the American commitment towards Taiwan. Chinese concessions on Taiwan tended to be made when more immediate strategic concerns took priority as they had done in 1972 and at the time of normalization in 1978. An indication of the new Chinese attitude was their failure to respond to the offers of access to arms sales and to establishing high-level military exchanges that Haig made in the course of his visit to Beijing in June 1981, even though the Chinese themselves had earlier requested them. Interestingly, the new communiqué of August 1982 involved a certain climb down by the Chinese side as it accepted that the Reagan administration intended to ignore its threats about a reversal of Sino–American relations and that it would proceed with the sale of arms to Taiwan without stipulating a date on which they would terminate.[5] The following month Hu Yaobang declared to the Party Congress that henceforth Chinese foreign policy was to be based upon independence. It seemed as if the period of basing Sino–American relations on a shared strategic opposition to the Soviet Union had run its course. Yet it was precisely from this point that relations in defence-related matters began to improve.

The period of commercial interaction, 1983–9

Although China's leaders professed dissatisfaction with the conditions of American arms sales to Taiwan and despite their disappointment at the relative downgrading of China's importance by Secretary of State George Shultz, they nevertheless took steps in conjunction with their American counterparts to address the remaining bones of contention in the relationship. So much so that by the autumn of 1983 Hu Yaobang described the quality of the relationship as very good (the meaning of the Chinese term 'not bad'). This was the first positive description of relations by a Chinese leader since Reagan took office more than two and a half years before. Since the agreements reached earlier in the year were over relatively minor issues the reason for the improvement in the relationship has tended to be attributed to the economic imperatives of China's renewed emphasis upon economic reform and rapid economic growth after a period of retrenchment.[6]

This is perhaps the first time that such an important shift in China's foreign policy can be directly associated with the needs of reform alone. This heralded the high point in Sino–American relations. Economic and cultural ties thrived. Trade grew from US$4.4 billion in 1983 to US$13.5 billion in 1988 and the sale of advanced technology to China increased from US$650 million in 1983 to US$1.7 billion in 1985, at which point it did not rise further in part because of the complexities of the American licensing system and in part because of the temporary raising of export controls in 1987 and 1988 as a result of American disquiet about Chinese arms sales to the Middle East. America was China's third largest trader and by 1988 it had become second only to Hong Kong as a source of direct foreign investment. The Chinese also appreciated the American role in encouraging China's admittance to a variety of the key international economic organizations and to facilitating the favourable treatment China received there. These not only provided soft loans for important projects under the control of the central government but also essential training for key Chinese personnel. The cultural and educational exchanges with the United States were also significant in this regard.

It was under these more relaxed conditions that military relations between the two sides flourished for the first time. Perhaps less was at stake for each side. The Chinese no longer feared Soviet aggression in the same way and indeed by 1985 Deng Xiaoping had stated in public that he anticipated that China would enjoy a peaceful international environment for the remainder of the century.[7] It meant that the Chinese were less exercised by the problems of becoming unduly dependent on the USA or of becoming strategically subordinate to it. The American side meanwhile had begun to redress the military balance with the Soviet Union through its own unilateral military build up. The Americans nevertheless still appreciated China's overall strategic role in holding down a significant proportion of Soviet forces in Asia. Under these conditions the two sides found it easier to cooperate either directly or indirectly over support for the insurgents in Afghanistan and Cambodia and in upholding the positions of their common allies of Pakistan and Thailand respectively. Although Sino–Soviet relations had begun to improve,[8] the Chinese were still concerned by the deployment of Soviet SS-20 missiles to the Far East, the stationing of TU-16 bombers in Cam Ranh Bay and by the enhanced Soviet support that enabled the Vietnamese to mount major offensives in Cambodia in the dry seasons of 1983–4 and 1984–5 that led to clashes with Thai troops on the border.

In 1984 China was made eligible for the Foreign Military Sales (FMS) programme, enabling it to purchase American arms with government financial assistance. High-level exchanges between the two military establishments led to agreement by the end of 1985 that the United States would be prepared to transfer weaponry in four defensive areas. Accordingly agreements were reached to coproduce anti-tank TOW line-guided missiles and on the export of American counter-artillery battery radar. The USA also agreed to assist in the production of large calibre fuses and detonators. Agreement was also reached to coproduce an anti-submarine torpedo that could be launched from surface vessels and to develop jointly an advanced avionics system for fifty of China's new F-8 fighters. These largely defensive projects were distributed between each of the People's Liberation Army main services. They spurred a rapid rise in American arms sales from US$8 million in 1984 to US$106.2 million in 1989.[9]

However the growing interdependence of China with the outside world and the deepening interconnections between Chinese and American institutions and societies were giving rise to new problems as the fundamental differences between the two sides became more evident. In the military field, the nationalist, neo-mercantalist orientation that still lay at the heart of much of the Chinese approach to the outside world became an issue in China's readiness to sell arms clandestinely to the Middle East and in American suspicions about China's possible proliferation of nuclear technology and other weapons of mass destruction.[10] These suspicions combined with Chinese refusals to allow for inspections, or to agree to establish what the Americans regarded as appropriate safeguards, stopped the Sino–American agreement on nuclear cooperation from going into effect. Moreover, by the late 1980s the new understandings reached between Reagan and Gorbachev combined to reduce China's global strategic significance. Already in the US Congress there were signs of concern about human rights questions in China, notably concerning Tibet after Chinese repressive acts there in 1987. But broadly speaking the general thrust of American policy was highly supportive of reforming China. To be sure matters were not helped by the tendency to misunderstand the reforms as leading to capitalism in economics and liberalism in politics, but it is likely that even if American popular perceptions had been more accurate they would still have reacted with horror to the killings of the peaceful demonstrators that they saw on television on the night of 3 and 4 June 1989.

The estrangement, 1989–94

The Tiananmen crisis in Sino–American relations brought to a head the fundamental problems in the hitherto incipient clash of values and approaches of the political cultures on which the PRC and the USA are established as states. And it posed the problem to the two sets of leaders as to how to balance that with their respective requirements to cooperate together in a number of areas of common interest. These problems were intensified by the collapse of the Communist regimes in Eastern Europe later that year and by the disintegration of the Soviet Union in 1991. These events intensified the fears of China's leaders about the survival of their regime and it encouraged Americans to believe that fundamental political change in China was just a matter of time.

Deng Xiaoping's response to these traumatic events reflected a certain underlying consistency in his approach. He has continually advocated that China needed to modernize by reforming its economy and opening up to the outside world, but it could only do so under the relative stability provided by unrelenting rule by the Communist Party. He made a point of reaffirming this approach in his address to his military commanders after the Tiananmen killings. In other words, even at the point of China's deepest domestic and international crisis when Deng was under pressure from his less reformist colleagues to draw up the socialist barricades against the capitalist world, he insisted upon adherence to his central strategy. While he argued that 'we should not have an iota of forgiveness for our enemies' he also declared that 'our reforms and opening up have not proceeded adequately enough'.[11] Deng upheld this view even after the collapse of Communism in Europe and the Soviet Union. While he recognized the dangers from the international 'enemies who never sleep and who will use every pretext to cause trouble, to create difficulties and pressures for us', Deng also called a halt to the leftist campaign against the dangers of 'peaceful evolution' lest the focus on alleged American conspiracies to undermine socialism by peaceful means should become a way of resisting further economic reforms and openness. Deng argued that what China needed was 'stability, stability and more stability'. In the course of his famous 1992 Spring Festival tour of southern China, Deng argued that there was nothing to fear from extending the operations of foreign companies in China: 'As long as we keep ourselves sober-minded, there is nothing to be feared. We will still hold superiority, because we have large and medium state-owned enterprises and township and town enterprises. More importantly, we hold the state power in our hands.'[12]

It was on this basis that the Chinese leaders pursued a more participatory role in the regional organizations of the Asia–Pacific and indeed of the international community as a whole. Thus to focus on military related questions, since 1989 China has signed the Nuclear Non Proliferation Treaty, promised to abide by the Missile Technology Control Regime, joined the Association of South East Asian Nations (ASEAN) Regional Forum and played a relatively constructive role in the UN Security Council. It has even accepted that a certain degree of international interchange on the domestic performance of human rights is legitimate. But it has drawn the line against external involvement in human rights issues deemed to be subversive of the Communist regime itself.

The American response has been far from consistent. President Bush announced a series of sanctions that had the merit of being selective and controllable by the executive branch of the US government. These principally suspended high-level exchanges, put an end to government and government-backed loans, and called a halt to military exchanges and military-related sales to China. Their renewal would depend upon satisfactory progress having been achieved in reforming the political system. Within two months Bush secretly dispatched his National Security Adviser and Deputy Secretary of State to Beijing where they met with China's most senior leaders. It has been claimed that these 'contacts' (not 'exchanges') did much to reassure China's leaders that America was not seeking to undermine their rule and that the visit together with other subsequent 'contacts' paved the way for them to pursue a more moderate approach at home and abroad in 1990.[13] Whatever may be thought of President Bush's policy of constructive engagement, its evident inconsistency with much of his public rhetoric enraged Congress and contributed to ending the domestic bi-partisan approach to China. Meanwhile the wisdom of indicating to the Chinese that American presidents did not mean what they said may be doubted.

The Congressional response by the summer of 1990 was to oppose the renewal of Most Favoured Nation (MFN) legislation which enabled Chinese goods to be imported into the United States under tariff conditions as low as any other country. The Congressional bill was vetoed by President Bush and although the House of Representatives easily mustered the necessary two thirds majority the same was not true of the Senate. The pattern was repeated the following two years leading to a situation in which the Chinese leaders looked to a more benign

President to defeat a more hostile Congress. The pattern was also set by which China's economic reforms were debated politically in the United States within a rather narrow and perhaps distorting frame of reference that centred on human rights and the significance of China's vast economic potential for American business. On human rights the issue seemed to be whether Bush's constructive engagement would encourage change through the emergence of an entrepreneurial middle class or whether democratic change would best be induced by sanctioning China's trade with the USA. The way the issues were debated and the options considered left little room for considering what kinds of reforms were best suited to American interests or what ranges of options were available in dealing with the Chinese government on reform matters while simultaneously seeking its cooperation on such questions as arms control and regional security.

The advent of President Clinton changed the American conduct of the MFN issue. With a Democratic majority in Congress, Clinton, in what was then praised as a 'superbly stage-managed' decision, extended China's MFN status in 1993 and made its renewal in 1994 conditional on the attainment of 'overall, significant progress' in seven nominated areas of human rights behaviour.[14] The administration expected that the Chinese government would make the appropriate adjustments in the ensuing twelve months. This perhaps reflected a long-standing American view as was evident in the approach of the Bush presidency that with the right combination of inducements and penalties the United States would be able to bring about fundamental changes in Chinese policies and character of governance.[15] It may be recalled that in the course of his confirmation hearings in the Senate Warren Christopher, the nominee for Secretary of State, explained that the administration's 'policy' was designed 'to facilitate a broad peaceful revolution in China from communism to democracy, by encouraging the forces of economic and political liberalization'.[16]

As 1993 unfolded Sino–American relations suffered. There was little evidence to show that the Chinese government had begun to treat the human rights of its citizens in accordance with American demands and the Chinese government tended to hold the Americans largely responsible for the failure of Beijing to be awarded the Olympic Games for the year 2000. Amid continual accusations that the Chinese had violated undertakings not to proliferate weapons of mass destruction and sell medium-range missiles, the United States tracked and eventually boarded a Chinese naval vessel which was claimed to be illicitly transporting chemical warfare materials to the Middle East. In the event none were found. At this stage the Clinton administration decided to halt the rapid deterioration of relations as these were damaging other priorities, notably the goal of establishing the United States at the heart of attempts to create an economic community of the countries of the Asia–Pacific. A series of high-level visits to China were initiated in the winter of 1993–4. President Clinton met President Jiang Zemin in Seattle at the Asian Pacific Economic Co-operation (APEC) summit. While the former addressed what he regarded as the dawning of an Asian Pacific community whose time was thought to have come, the latter took the opportunity to lecture the American about Chinese conceptions of human rights.

From spring 1994 Clinton's China policy began to unravel. Within a context in which the American administration had antogonized many of the East Asian governments, including the normally well-disposed ones such as Indonesia, Singapore and Japan, the prospects for an Asian Pacific economic community that were always going to be difficult to realise began to recede very rapidly even though it was only in November in Seattle that President Clinton had put it forward as a major objective. Under pressure from American exporters and desirous of Chinese assistance in handling the North Korean question, Clinton not unexpectedly gave way on the MFN question even though he acknowledged that the Chinese had not made the progress on human rights for which he had called.

President Clinton's conduct of policy in East Asia reached such a point that his leading official on Asian affairs, Assistant Secretary of State Winston Lord, wrote a damning indictment of it to his Secretary of State Warren Christopher, warning that the country's relations with Asia were being derailed by a 'malaise of disputes with several key countries over human rights, trade and other matters'.[17] The critique has been elaborated upon by one of Washington's leading academic authorities on Asian affairs who lays much of the blame on the failure of the President to provide proper leadership and on the inadequacies of the bureaucratic and political processes in Washington.[18] This suggests that the administration is unlikely to pursue any new substantive initiatives until it undertakes a major overhaul of its decision-making processes. But it is not clear that the absence of effective American leadership at this point has caused fundamental damage to American enduring interests.

THE DEFENCE DIMENSION

Given the amount of attention that successive administrations have devoted to the issue, actual American arms sales to China have been of minimal commercial value. As we have seen, even during the halcyon days of Sino–American interactions from 1984 to 1989 when arms sales were agreed after careful consideration by both sides, they barely exceeded $100 million in total. The American contribution to China's defences, however, should not be judged by these figures, but rather by the less easily calculable factors of technology transfers and by the familiarization of the Chinese military with modern knowhow. Indeed much of this increased after 1989. In 1990 US exports to China were composed proportionately in the following percentages: agricultural products 11, chemicals 19, other raw materials 23, heavy machinery 35, electronics 5, and other manufactures 7. By 1992 US exports had changed: aircraft and parts were dominant followed in order of magnitude by computers, generators, fertilizer, telecommunications, instruments, cereals, vehicles, chemicals, plastics, petroleum, and cotton.[19]

Interestingly, the sanctions against arms sales to China were less severe than may have been thought as they applied only to items contracted under the US foreign military sales (FMS) rules. Among those that involved direct commercial transactions which missed the FMS net were deals to upgrade China's F-7M fighters and the F-8 air defence interceptor. President Bush waived sanctions to allow a US-made satellite to be launched by a Chinese rocket and to allow the sale of four Boeing 757s whose inertial reference systems were on the munitions control list. But some of these deals were then cancelled by the Chinese side – less because of errors of omission or commission by the American side (as tended to be claimed in Washington) than because of Chinese purchases of Soviet aircraft and attempts to produce others under licence.[20] The Clinton administration has also waived sanctions on several items of so-called dual-use technology.

The exchange of military personnel and the participation of Chinese military officers in joint seminars, tours of American military installations, inspections of advanced weaponry and the general establishment of personal contacts and networks were affected by the sanctions. This did not diminish the high regard with which the Chinese officer corps held American military capabilities. Any doubts were immediately dispelled by the success of American and West European forces in the Gulf War where their superior technology easily overcame Iraqi forces who were equipped to a standard generally higher than that of the Chinese themselves. But it is not only the Chinese side that may regret the prolonged absence of personal contact. One of the few concrete achievements of Warren Christopher's visit to Beijing in March 1994 was to reach an agreement to restore military exchanges and set up a joint commission for defence conversion.[21] There can be little doubt that in the long run such contacts will facilitate better understanding, but it should not be thought that in themselves such contacts would stop significant members of the Chinese military from believing that the United States is their country's long-term enemy.[22]

Meanwhile it remains true that the American primary contribution to China's defence efforts in the current period of economic reform is indirect through its contribution to the Chinese economy. In addition to China's enormous trading surplus that has climbed steadily from US$1.5 billion in 1989 to US$25 billion in 1993, US exports have also risen in value to reach US$8.8 billion. The USA has become the largest direct investor in China (after the exceptional case of Hong Kong) with committed investments of US$11 billion. These investments also tend to be in large scale projects and to make use of 'more high technology'.[23]

THE CULTURE OF FOREIGN POLICIES

Since the ending of the Cold War the underlying cultural and political differences between China and the United States have emerged to cast a pall over their relationship. These were kept largely at bay during the period when they shared an opposition to the Soviet Union. Although it was an uneasy partnership that flourished only in the mid-1980s when the two felt less dependent upon each other, the strategic priorities generally prevailed over the differences that divided them. However, the conjunction of the Tiananmen events with the collapse of Communism elsewhere and the disappearance of the strategic framework that gave them a common purpose has subjected the relationship to stress and fragmentation. Ironically, it is one that has also flourished in many ways. Trade has grown rapidly, as have many aspects of an increasingly complex economic relationship. After a decent interval cultural, scholarly exchanges have resumed as has tourism.

The American contribution to Chinese efforts to upgrade the technological levels of their defences in the course of economic reforms has been greatly affected by these broader developments. This has been particularly evident since 1989 when military exchanges between the two sides were reduced to a minimum. The visit by Warren Christopher in March 1994 paved the way for their renewal. But other considerations have also been relevant. On the American side there has been concern about the possible impact on third parties of arms sales and transfers of dual-use technology to China. In the Carter era and the early years of the Reagan administration these centred on their possible impact on the Soviet Union and the prospects for Sino–Soviet détente within the context of the 'strategic triangle' and they gave rise to significant divisions between senior policy-makers which were not settled until the appointment of Shultz as Secretary of State and the reassertion of American power due to Reagan's rearmament policies.[24] Another concern was how the upgrading of China's military capabilities may affect Taiwan's capacity to defend itself. A further issue was China's own arms sales that were deemed to be irresponsible and fears about China's possible proliferation of weapons of mass destruction. At the same time the fact that China was still a Communist country affected American policy in ways which were not particularly consistent, but were nevertheless evident from time to time. Indeed during the Bush administration arms sales were regarded as part of the policy of constructive engagement designed to cause the Chinese government to change some of its Communist ways regarded as most objectionable in America.[25] To be sure there were also the long-standing arguments of a more straightforward commercial kind that arms sales were good business and good for business and if America did not sell others would. But American caution and debates about arms sales reflect the other concerns. Indeed any future American arms sales to China will doubtless be constrained by the contemporary strategic environment in the Asia–Pacific in which China is no longer threatened by Russia and in which American allies in the region may be adversely affected by the rise of Chinese power.

It is important to recognize that the Chinese side has also been subject to constraints in developing defence relations with the United States. Despite the high regard for American advanced military technology evident in China, shortages of funds, a desire to acquire production capacities rather than purchase weapons systems off the shelf, and China's own technological backwardness have combined to limit actual purchases of foreign arms in general. But purchases from the United States have also been constrained by political factors. These range from fear of being manipulated by the superior power and of not being treated on equal terms to the use of military exchanges (or rather the withholding of them) as one of the levers in the conduct of diplomacy.[26] It should also be noted that the Chinese have acquired advanced weapons systems and hightech military equipment from American allies such as Israel, France and Britain. More recently these have been acquired from Russia. Nevertheless the Chinese have evinced great interest in gaining access to advanced technology and this includes military-related technology from the United States which is regarded as a key source for both.

America's role in the modernization of China's defences must be understood within the context of Sino–American relations as a whole.

As these are a complex mixture of conflictual and cooperative interests the issue turns on the capacities of the two sides to manage the relationship. Once again we return to the problems of the differences of culture and politics that divide the two sides and how to ensure that these do not undermine this important relationship even though it may be impossible to escape their influence.

The difficulties on the American side are all too apparent. A pluralistic system in which policy-making, though centred on the President, is diffused within often competing bureaucracies and to an extent shared with an often recalcitrant Congress has not been placed to define and pursue a coherent and coordinated foreign policy in the post-Cold-War period. These inherent systemic problems have been compounded by a President who has paid only intermittent attention to foreign policy. Indeed, as noted earlier, his own most senior official on Asian affairs has complained about a current 'malaise' in American policy. Yet the way in which the MFN issue has been subordinated to human rights for which the Clinton administration may be faulted is also a product of the American political system itself, involving as it did an agenda derived from the stand off between the previous President and Congress. That agenda made little attempt to distinguish between those aspects of human rights that confronted the Chinese system and those that could be said to push the reform process. For example, it is easier for the Chinese authorities to agree on terms of emigration or even to ban exports of prison-made goods than it is to free all political dissidents. Little consideration was given to demanding that the Chinese authorities follow those aspects to which they are bound by their own signatures to international agreements or even by their own laws regarding the procedures for arrest, treatment and detention of all citizens. The American agenda was clearly derived more from domestic sources which could only have the effect of accentuating the cultural divide.

The Chinese side, paradoxically may be said to have the greater problems. Since political power is more centralized, Beijing has been able to follow a more coherent and consistent policy. But as the troubles of succession loom the country is faced with immense social and economic problems that its political system is ill equipped to manage. Such a situation invites caution and a particular truculent rigidity in approaching the outside world and America in particular. For all its faults the American system is both durable and flexible – attributes that cannot be associated with China's Communist system despite its enormous successes in economic reform.

NOTES

1 Harry Harding, *A Fragile Relationship: The United States and China since 1972*, Washington, DC: The Brookings Institution, 1992.
2 Steven I. Levine, 'Sino–American Relations: Testing the Limits of Discord', in Samuel S. Kim (ed.) *China and the World: Chinese Foreign Relations in the Post Cold-War Era*, Boulder, Colorado: Westview Press, 3rd Edition 1994, p. 90–91.
3 For details see Gerald Segal, *China Changes Shape: Regionalism and Foreign Policy*, Adelphi Paper no. 287, London: International Institute for Strategic Studies, 1994.
4 Except where stated this account draws principally upon Harding, op. cit.
5 See Jaw-ling Joanne Chang, 'Negotiation of the 17 August 1982 U.S.–PRC Arms Communique: Beijing's Negotiating Tactics', *The China Quarterly*, no. 125, March 1991, pp. 33–54; and Robert S. Ross, 'China Learns to Compromise: Change in US–China Relations, 1982–1984', ibid., no. 128, December 1991, pp. 742–73.
6 See the argument by Harding, op. cit., pp. 139–40.
7 Deng Xiaoping, *Fundamental Issues in Present Day China*, Beijing: Foreign Languages Press, 1987, p. 116.
8 For details see Gerald Segal, *Sino–Soviet Relations After Mao*, Adelphi Paper no. 202, London: International Institute for Strategic Studies, 1985.
9 See Harding, op. cit., pp. 168–9.
10 For an account of Chinese neo-mercantalist attitudes see Samuel S. Kim, *China In and Out of the Changing World Order*, Princeton: World Order Studies Program Occasional Paper no. 21, Princeton University, 1991.
11 See Deng Xiaoping's speech of 9 June 1989 in *Beijing Review*, vol. 32, no. 28, 10–16 July 1989, pp. 14–17.
12 For a more extended treatment of Deng's arguments and for sources of his quotations see Michael Yahuda, 'Deng Xiaoping: The Statesman', *The China Quarterly*, no. 135, September 1993, pp. 551–72.
13 David Zweig, 'Sino–American Relations and Human Rights' in William T. Tow, *Building Sino–American Relations*, New York: Paragon House, 1991, pp. 72–3.
14 Susumu Awanohara, 'Breathing Space, Clinton delays on conditions to China's MFN renewal', *Far Eastern Economic Review*, 10 June 1993, p. 13.
15 For detailed criticisms along these lines of the China policy of the Bush administration see June Dreyer, 'Military Relations: Sanctions or Rapprochemenent?' in Tow (ed.), *Building Sino–American . . .* , op. cit., pp. 203–20.
16 Cited by Daniel Williams and Clay Chandler, 'As Stand on China Collapses, Clinton Tries to Save Face', *International Herald Tribune*, 13 May 1994.
17 Daniel Williams and Clay Chandler, 'Senior State Dept. Aide Warns White House of Pitfalls in Asia Policy', *International Herald Tribune*, 6 May 1994.
18 Harry Harding, 'Asia Policy to the Brink', *Foreign Policy*, no. 96, Fall 1994, pp. 57–74.
19 Harding, op. cit., pp. 365–6.
20 For details and discussion see Dreyer, 'Military Relations. . .', op. cit., pp. 205–8 and p. 214.

21 *Beijing Review*, 21–27 March 1994, p. 39.
22 Paul H. B. Godwin, 'Force and Diplomacy: Chinese Security Policy in the Post-Cold War Era', in Kim (ed.), *China and the World*, op. cit.; see also, David Shambaugh, 'Growing Strong: China's Challenge to Asian Security', *Survival*, forthcoming.
23 Wu Yi, Minister of Foreign Trade and Economic Cooperation, after her visit to the China–USA trade fair in Los Angeles on 11 April 1994. The fair led to the signing of contracts to the value of US$1.3 billion and to 134 agreements valued at US$4 billion. *Beijing Review*, 2–8 May 1994, p. 6.
24 Banning N. Garrett and Bonnie S. Glaser, 'From Nixon to Reagan: China's Changing Role in American Strategy', in Kenneth A. Oye *et al.* (eds), *Eagle Resurgent?: The Reagan Era in American Foreign Policy*, Boston: Little Brown, 1987.
25 See Dreyer, op. cit., p. 209.
26 For example, the Chinese government stopped all military exchanges between 1981 and 1983. See Harding, op. cit., pp. 134–43.

8 China and East Asia

Masashi Nishihara

In the last few years China has made a remarkable political and economic comeback after the fiasco of the 1989 Tiananmen Square incident. All major countries, particularly East Asian, now court China. China's gross national product grew by 12 per cent in 1992 and 13 per cent in 1993. The attractiveness of the huge Chinese market, which has enormous potential, has given the Beijing leadership the sense that their country's international standing has been enhanced.[1]

What has been the impact of this economic self-confidence upon China's security policy toward East Asia?

Despite the numerous vulnerabilities of China's economic reform such as widespread corruption, growing economic discrepancy between urban and rural sectors, soaring financial deficits of the central government, the current Beijing leadership appears to be determined to adhere to the policy of reforms and opening up to the outside world, since 'this is the only way to build a modern socialist China, and transform our motherland into a strong, prosperous, democratic and civilized country'.[2]

China wants to be strong and prosperous. To achieve this goal, Beijing needs a peaceful environment. However, in order to create such an environment, China seems to be seeking two apparently contradictory paths at the same time: to participate in multilateral economic and security arrangements such as the Pacific Economic Co-operation Council (PECC), Asia-Pacific Economic Co-operation (APEC), Association of South East Asian Nations (ASEAN) Regional Forum, and at the same time to increase its military power, establishing a military dominant position in Asia by taking advantage of cuts in the American and Russian armed forces and defence budgets.

SELF-CONFIDENCE IN DIPLOMACY AS A STEP TO HEGEMONISM?

As China's economy has grown, it has also gained diplomatic self-confidence. In July 1993 the ASEAN Post-Ministerial Conference, establishing the new ASEAN Regional Forum, accepted China as a full member. In November, when President Jiang Zemin went to Seattle to participate in the APEC Leadership Conference, he managed to prevent the attendance of Taiwan's counterpart, although Taiwan is also a full member of APEC.

China's diplomatic self-confidence is now accompanied by a degree of arrogance. In September 1993 China lost the hard-fought bid for the Olympic Games in the year 2000, and soon after it conducted a nuclear test. In May 1994 China succeeded in having Washington decouple the extension of most favoured nation (MFN) from the human rights issue, and soon after it tested a nuclear device. These well-timed actions by Beijing seemed retaliatory in the first case and defiant in the second. China, in fact, defiantly arrested political dissidents before the American Secretary of State visited Beijing in March 1994. His visit was timed with the convening of the People's Congress. China may claim that this is a standard procedure prior to any large political meeting, but the Beijing leaders must have calculated the impact which the arrest of political dissidents at the time of an American leader's visit might have upon American recognition of emerging Chinese power.

Where will this diplomatic self-confidence lead? Will China become an active participant in multilateral security consultations, or will it begin to act more as a self-assertive hegemonic power, establishing a new regional order under 'the Middle Kingdom'?

CHINA AND RUSSIA

Current relations between China and Russia, rivals during much of the Cold War, are on the whole friendly. President Yeltsin visited Beijing in December 1992. In August 1993 three naval ships of the Russian Pacific Fleet visited Quindao, Shandong. It was the first port visit to China by Russians since Tsarist times. In the same month China's Chief of Staff Zhang Wannian visited Moscow, and agreed with Russian Defence Minister Pavel Grachev to promote military cooperation. The visits of both Yeltsin and Defence Minister Grachev produced military cooperation agreements, including the transfer of 26 SU-27 fighters.

For Russia, China has become the second largest trading partner after Germany. In 1993 the Chinese–Russian trade soared to US$7.7 billion,

30 per cent more than in the previous year. Russian Foreign Minister Andrei Kozyrev visited Beijing in January this year and discussed further economic cooperation. Chinese Foreign Minister Qian Qichen visited Moscow in late June, and President Jiang Zemin in September. The leaders of the two countries stressed their friendship and signed several agreements for mutual cooperation, including security co-operation, while exchanging views on many international issues, particularly the Korean problem.

While Russians, who are short of daily goods, obtain great benefits from trading with the Chinese, their growing economic dependence upon China worries their security specialists. Border towns such as Manzhouli are prospering.[3] For Russia, China's economic might and explosive demographic conditions are potential sources of threat to security.

There are two potential sources of conflict between the two countries. First, tensions are growing in Russia's far east, as more Chinese workers and traders move into major cities such as Vladivostok and Khabarovsk. Most estimate the number of legal and illegal Chinese in the Russian far east at approximately 300,000, but some sources even quote as many as one million.[4] Some take over jobs and markets from Russians, causing serious frictions with local communities. Second, the two countries still have not concluded border demarcation talks along the Amur River. This is in part a reflection of Chinese nationalists. Some Chinese consider Russia a long-term threat. Some even go so far as to claim Russia's far eastern region as their territory, referring to the Russian annexation in the nineteenth century.

China's interest in developing the Tumen River delta promotes its desire to conduct trading activities through the Sea of Japan, but China may also be trying to establish a naval presence there. In 1984 for the first time the Chinese Navy sent three training ships into the Sea of Japan. In 1991 China sent a naval oceanographic observation ship there for a month, which was interpreted as a step towards establishing China's naval presence.[5]

CHINA AND MONGOLIA

Mongolia's post-Cold-War foreign policy is to maintain equidistance with its two big neighbours, Russia and China, and to increase contacts with the developed world, particularly the United States and Japan. Maintaining equidistance with its former protector and its former adversary is an extremely difficult problem. Mongolian–Chinese trade has sharply increased in the last few years. Exports to China went up from 12 per cent to 25.7 per cent of Mongolia's total. Imports were 20.7

per cent in the first half of 1993 trade. Current trade relations show that Mongolia depends upon Russia for 50 per cent of its trade and upon China for 25 per cent of its trade.[6] Some Mongolians argue that they should increase trade relations with China even more so that it can balance Russia.

The traditional fear that China may dominate Mongolia is reinforcing the current re-emergence of Mongolian nationalism, as is shown by the intense interest in reviving the admiration of Genghis Khan and identifying the location of his thirteenth-century capital.[7] Being apprehensive about the closer relations that may develop between Mongolians in Mongolia and Inner Mongolia, the Beijing government has placed restrictions on the movement of people between the two regions since 1992. Perhaps to balance Chinese clout, President Ochibrbat visited Moscow in January this year for a meeting with President Yeltsin and to sign a bilateral treaty of economic cooperation.[8]

CHINA AND THE TWO KOREAS

China lost control over the Korean peninsula to the Japanese in 1895. During the Cold War China managed to keep the northern half under its sphere of influence, although it often had to compete with the Soviet Union. Only in August 1992, two years after Russia recognized Seoul, did China establish diplomatic relations with the southern half. Beijing has since increased its influence in South Korea, surpassing that of Russia.

In the postwar period China maintained its special status in North Korea by dispatching a sizeable number of 'volunteers' in support of North Korea during the Korean War, by having a treaty of mutual assistance of 1961, and by providing substantial economic and military assistance. North Korea was not always a loyal partner for China, however, because of this historical relationship, and as a large neighbour sharing the border with North Korea, China has had a strong hand in that country. In 1975 when the fall of Saigon led Kim Il Sung to call for the liberation of the south, China, which wanted to have a strategic relationship with the United States, strongly warned Kim not to take military actions against South Korea.

Since the late 1970s China has apparently tried to persuade the North Korean leaders to open their economy. Beijing, also, supplies nearly half of Pyongyang's oil needs and thus can influence North Korea in many ways.

When North Korea found itself in international isolation over its alleged plan to develop nuclear arms, China was Pyongyang's only ally.

China opposed any economic and military sanctions against North Korea either by the UN Security Council or by the United States and its allies. This opposition made China's hands even stronger. In mid-June 1994 when North Korea's Chief of General Staff visited Beijing, apparently to seek China's support for Pyongyang's 'nuclear diplomacy', Admiral Liu Huaquing, vice-chairman of the Party's Military Commission, gave him a stern warning against any provocations by the north.[9] China does not want to have any armed conflict in the peninsula, which is likely to jeopardize its own economic growth, and is afraid of a massive flood of refugees across its border.

Since diplomatic relations were established between Beijing and Seoul in August 1992, their trade volume has increased rapidly, with China having a trade surplus. The volume of trade between the two countries in 1993 reached US$10 billion, which was 40 per cent higher than the previous year. China became the third largest trading partner in 1993, and is expected to become the second largest trading partner in the near future.[10] Diplomatically as well, Beijing is in an advantageous position as Seoul had to court Beijing's influence over Pyongyang with regard to nuclear issues. South Korean President Kim Young Sam has visited Beijing twice in the last two years, whereas President Jiang Zemin has not been to Seoul.

China has managed to establish official relations with South Korea without jeopardizing its relations with North Korea. South Korea, in this sense, has miscalculated. It apparently thought that China's formal relations with economically stronger South Korea would weaken its relations with North Korea.[11] Instead, China has annoyed Seoul by dispatching its naval oceanographic research ships to the Yellow Sea in April 1993 to explore offshore resources.[12] In 1991, South Korea had to stop its similar activities because of China's strong protest. This high-handed behaviour by China accords with the nature of traditional relations between China and the Korean peninsula.

The Russians are also competing with the Chinese to establish influence in the peninsula. Moscow's recognition of South Korea in 1990 and the demise of the Soviet Communist Party in 1991 cooled relations between Moscow and Pyongyang. Russia's efforts to regain its influence over North and South Korea are seen in its proposal for a six-power conference over North Korean nuclear issues. North Korea showed no particular interest, since its primary strategy seemed to be to enter separate bilateral talks with the USA and South Korea; neither did South Korea.

During President Kim Young Sam's visit to Moscow in early June 1994, President Yeltsin tried to strengthen his hand by showing

Russia's support of the UN Security Council's possible sanctions (in exchange for South Korea's resumption of committed credit as well as Seoul's push for Russia's faster entry into the Asia–Pacific economy). Thus both North and South Korea are locked into a complex relationship with both Beijing and Moscow.

CHINA, JAPAN, AND THE EAST CHINA SEA

Despite the fact that China and Japan maintain close relations in virtually all fields on the surface, theirs is essentially a fragile relationship. Emperor Akihito visited China in 1992, which helped to forge relations between the two countries. However, the Chinese remain concerned about the rise of Japan as a political power that now sends its troops overseas in UN peacekeeping operations and is seeking a permanent seat on the UN Security Council. The Japanese, on the other hand, are becoming apprehensive about China's military build-up, particularly naval build-up, its continued supply of missiles to areas of potential conflict, and its nuclear tests.

China fears that if 'economic contradictions' should develop more seriously between Japan and the United States in the future Japan may seek a foreign and defence policy independent of Washington. China appears to consider Japan as a major source of threat in the early twenty-first century. The Beijing government has been cautious about supporting Japanese aspirations to attain a permanent seat on the UN Security Council, for it would weaken its own influence in regional and global politics. This suggests that there is a potential possibility of Sino–Japanese rivalry.

Japan tries to check China's military build-up by considering the possibility of reducing its official development aid to China and by proposing bilateral security dialogues in order to enhance the level of military transparency.[13] Japanese aid is now more carefully disbursed so that it will not be used for infrastructure projects that have military implications. More aid is being given to environmental improvement programmes. The first bilateral security dialogue took place between the foreign ministries in December 1993, while the second one, this time between the defence ministries, was held in Beijing in March 1994. In addition, in January 1994, Foreign Minister Tsutomu Hata raised Japan's concern with growing Chinese military power.

Japan knows that China is vulnerable to international criticism of its lack of military transparency. Tokyo has urged China to comply with the Missile Technology Control Regime (MTCR), the UN arms registry and to issue a defence white paper. Meanwhile, China knows that the Japanese are vulnerable to its condemnation of their wartime conduct.

China's offshore oil exploration, its harassing of foreign commercial vessels on the high seas, and the growing number of illegal Chinese migrants are more recent sources of annoyance to Tokyo. In the East China Sea some Chinese access to oil exploration is said to cross the middle line between Japan and China.[14]

Foreign vessels have been intercepted on the high seas in the East China Sea by what look like official Chinese vessels, which do not carry national flags but whose crew speak Chinese. Foreign vessels, including Japanese, were boarded, fired upon, or attacked. Some goods on the vessels were also confiscated. It certainly looks a lot like piracy. According to the report of the International Maritime Bureau in Kuala Lumpur, the incidents for 1992–3 included 20 in the East China Sea, 11 in the South China Sea, and 33 in the Hong Kong-Luzon-Hainan zone.[15] Chinese authorities apparently defend such operations as legitimate, by referring to the need to control smuggled goods such as cigarettes. In July 1993 the Russian Pacific Fleet was exasperated by such 'piracy', and dispatched several frigates in the East China Sea to protect its own commercial vessels and to threaten retaliation.[16]

Illegal Chinese migrants often force themselves into Japanese shores during the night. The fact that those smugglers have increased in number reflects the decline of the central government's authority to control large operations by smuggling rings, which often implicate Japanese yakuzas. This does not yet pose a security threat to Japan, but is raising a new concern among security specialists that in the case of domestic turmoil in China a much larger scale of smugglers or refugees may cause internal security problems.

CHINA, TAIWAN, AND HONG KONG

Chinese–Taiwan and Chinese–Hong Kong relations cannot be seen in the same context as China's relations with the rest of East Asian countries. They have unique historical backgrounds: Taiwan being governed by the Kuomintang, which used to rule the mainland, and Hong Kong being a British colony. Yet the common theme of this chapter that Beijing wants to exert influence over its neighbouring areas holds true here, too. There is some difference between Taiwan and Hong Kong, however: Taiwan, being geographically separated from the mainland, is in a stronger position *vis-à-vis* the mainland than Hong Kong.

Relations between China and Taiwan show both reconciliatory and emulative signs. They are developing close economic relations, to the extent that in April last year top leaders of non-governmental exchange organizations from both sides met in Singapore. It was the first such meeting, being indicative of the progress toward pragmatic cohabitation across the straits. At the same time, however, Taiwan aggressively pursues a new strategy to seek more formal contact with Southeast Asian countries. Late in 1993 and early in 1994 Premier Lien Chan visited Singapore and Kuala Lumpur, while in February 1994 President Lee Teng-Hui visited Manila, Jakarta, and Bangkok. Both of them, announcing their journeys as 'a holiday trip', met top leaders in each capital.

Because of Taiwan's great economic power and higher standard of living, Beijing cannot control its external policy. Yet, as China's economic power grows, it can acquire capabilities to project its strength. Its missiles, aircraft, larger ships, and the like can reach Taiwan. This will give Beijing a wider range of military options to 'liberate' Taiwan. While Taiwan's acquisition of 150 F-16s and 60 Mirage-2000s will help balance China's efforts in developing power projection capabilities, China may eventually possess military capabilities effectively to attack Taiwan. Taiwan then may also possess longer range missiles to counter them. The arms race across the strait is likely to continue.

For many years Hong Kong stimulated China's economic growth. Today China's economy now stimulates Hong Kong. In recent years the conflicts between Beijing and Governor Chris Patten over the democratization of Hong Kong elections have worried many observers. Yet the local economy has not been affected by the political conflicts.

Nonetheless, China already has a plan to introduce 10,000 troops for the defence of Hong Kong. In June 1994 China and Hong Kong reached an agreement in which the British will transfer fourteen out of thirty-nine military sites to the Chinese People's Liberation Army. The British are closing the large Tamar Naval Base, partly to prevent the Chinese in the post-reversion era from using it for its own naval activities. The Chinese have said that they would use the military facilities which would be turned over to them, only for defensive purposes. This remains to be seen.

Beijing's interest must be to turn post-1997 Hong Kong into something like Singapore where economic freedom is fully accepted while political freedom may be restricted. Governor Patten is opposed to such a strategy. Political tensions are thus likely to continue, with China always threatening to dismantle any 'democratic' measures that the British might introduce before 1997 and which are not written in the Basic Law. On the whole China holds an advantage over Britain.

CHINA AND INDOCHINA

China's relations with Southeast Asia continue to be a mixture of mutual benefit and suspicion. The fast growth of the Chinese economy attracts Southeast Asians for trade and investment. At the same time, Beijing's continuing military build-up and its naval activities in the South China Sea and the Indian Ocean generate suspicion about its motives.

Historically China has held a dominant position over Indochina, intervening in internal affairs and supporting whatever forces they have found to be promoting their own geostrategic interests. Its policy to assist the anti-Hanoi Khmer Rouge in Cambodia and to 'bleed white' anti-Chinese Vietnam made the peaceful settlement of Cambodia impossible. Therefore, the change in China's policy toward the Cambodian peace process positively affected its relations with Vietnam. When China as a permanent member of the UN Security Council and as one major player in the Cambodian issue decided to withdraw its support of the Khmer Rouge forces and endorse the provisional UN presence, the peace process started to make quick progress and brought about the Paris peace agreement in October 1991.

However, in this process China demonstrated once again its high-handed posture toward its southern neighbour. In September 1990, it is now widely believed, China and Vietnam had a secret summit meeting in Chengdu City, southern China, with Jiang Zemin, Yang Shangkun, and Li Peng from the host side and Phan Van Dong, Nguyen Van Lin, and Do Muoi from Hanoi. Reportedly, the Vietnamese side was pressured to accept the UN peace framework, adopted by the Security Council in the previous month, and to agree to oust Foreign Minister Nguyen Co Tach, known to be an anti-Chinese leader. It must have been a humiliating meeting for Vietnam.

Soon after the Paris peace agreement was reached in October 1991, Party Secretary Do Muoi and Premier Vo Van Kiet visited Beijing to normalize relations between the two countries. China returned the visit, not by sending the party secretary but only Premier Li Peng, in September the following year. This traditional pattern of top Vietnamese leaders visiting Beijing remains unchanged. China holds the upper hand over Vietnam.

Even after the new Sihanouk government was established by the UN-sponsored elections in May 1993, China continued to retain influence over Cambodian politics. King Sihanouk lives in Beijing, and China offers a meeting place for other Cambodian leaders such as Prince Norodom Ranariddh and Hun Sen to consult with the King. In October 1993 China and Vietnam signed a border agreement, but failed to reach an agreement on the territorial waters. Although both sides agreed on the peaceful settlement of disputed islands and China stressed the joint development while postponing a final settlement, they continue to build military and other installations. Thus tensions continue in the South China Sea, with China having superior arms. The SU-27 fighters which China has purchased from Russia will probably be deployed in Hainan, making it possible for Beijing to have control over the airspace of the Paracels and Spratly islands. Vietnam is today in an inferior position, with seriously inadequate naval and air power.

CHINA AND ASEAN COUNTRIES

The traditional pattern of Chinese neighbours sending high-ranking officials to Beijing remains in current ASEAN–Chinese relations. During 1993 Prime Minister Goh Chok Tong of Singapore (April), President Fidel Ramos of the Philippines (April), Prime Minister Mahathir of Malaysia (June), Prime Minister Chuan Leekpai of Thailand (August), and King Bolkiah of Brunei (November) visited Beijing. China has returned these visits only with a leader of more than one rank below.

The above-cited list of ASEAN leaders' visits to Beijing demonstrates the attraction of those countries to the Chinese market. As their economic interdependence with China increases, they tend to emphasize the importance of 'Asian solidarity', at the expense of their relations with Washington. Recent Chinese–Malaysian relations are typical of this mood.

The Malaysian prime minister favours close relations with China, not only for economic reasons but for strategic reasons. He wants to have close security relations with Beijing, since the United States has closed its bases in the Philippines, thus creating a power vacuum in Southeast Asia. Mahathir also seems to want to balance Japanese economic power by drawing in Chinese economic power or by encouraging close ties between Malaysians of Chinese origin and China. In August 1992 Malaysian Defence Minister Razak was known to have visited China to promote cooperative relations between the two armed forces. In May 1993, it was reported that Chinese Defence Minister Chi Haotian went to Kuala Lumpur to discuss the possibility of training Malaysian forces in China as well as selling military equipment to Malaysia. In 1991 the two governments also agreed that China would launch a first satellite for Malaysia before 1995. In May this year Chi Haotian returned to Kuala Lumpur again.

Mahathir's strategy is apparently to resist the economic blocs of Europe and North America by starting to organize an East Asian Economic Caucus (EAEC). For this, China's and Japan's support are essential. Since Tokyo has been reluctant to give full support to the concept to the extent that it is designed to rival NAFTA, Malaysia tends to count more on Beijing. In June 1994 Mahathir headed a 300-member delegation to China.

While maintaining security arrangements with Washington, Thailand has had longer strategic relations with China than Malaysia. Bangkok has needed Beijing in order to balance Vietnamese power. Bangkok has bought a substantial amount of military hardware from China, partly because their prices are reasonable. The Royal Thai Army has tanks, armoured personnel vehicles, and artillery from China, while the Navy has a few frigates. As China increases its economic power, its defence industry, which is likely to become stronger, will tend to promote arms exports. This will forge stronger military relations between China and Thailand.

China also retains close relations with Myanmar (Burma). Its large-scale supply of arms and ammunitions to the current military regime (SLORC) is being suspected by some ASEAN nations and India as its strategic move to establish access to the Indian Ocean. The military supplies, which include fighters, high-speed attack boats, rocket launchers, tanks, and trucks, are said to be valued at US$1.2 billion. Chinese engineers have been sent to repair three naval bases and construct two new bases in small islands in the Indian Ocean. Reportedly, a naval base in one of the islands in the Bay of Bengal has installed new Chinese-made radar, which may be operated by Chinese engineers.[17] Here, too, China seems to have a strong hold over Myanmar.

THE CASE FOR CHINA'S ECONOMIC SLOWDOWN

Assessing Chinese–East Asian relations in recent years, one can detect a common pattern of Chinese external behaviour – one that seeks to secure a superior position over neighbours, which is observed in the asymmetrical levels of visitors to and from China. China tends to send the lower rank of government leader to the region than that of the East Asian leaders who visit China. Its strategic interest is to build a strong and prosperous nation in order to be able to assert its own interests in relations with its neighbours.

Beijing appears to find it easier to handle Mongolia, North Korea, Vietnam, Cambodia, Laos, and Myanmar than to handle other countries in the region. They are economically and militarily weaker than China. Beijing can intimidate them with economic and military sanctions, as it has done in the past. Where its relations with many of the other wealthier or stronger East Asian countries are concerned, Beijing cultivates cooperative ties with them in the hope of driving economic and political wedges between them and the United States. This does not mean that China seeks no cooperative relations with the USA, for it desires to build better relations with Washington. However, it also wants to minimize the possibility of having to compromise its own economic and security interests.

The desire to secure a superior position over its neighbours has been reinforced by China's remarkable economic success and by the absence of a clear, united protest or resistance by China's East Asian neighbours. Such desire has strengthened its political self-confidence. This has tended to make China behave more like a self-assertive player. China, while advocating joint developments in the South China Sea, for instance, shows no willingness to compromise on its sovereign rights over the islands. But the region does not need a strong China.[18] The evolution of the Chinese navy from a defensive coastal force to an offensive blue-water mission would destabilize the region and change the balance of power.[19] The trend will be accelerated as China's economy develops.

Naturally, Beijing's economic future is uncertain. If the 'socialist market economy' or 'Deng Xiaoping-style capitalism' should run into serious difficulty with, for instance, the rise of regionalist tendencies, the Chinese leadership will probably have to slow down the pace of developing power projection capabilities. This would be more acceptable to China's neighbours.[20] China cannot expand national power at the expense of economic welfare. In order to ensure economic welfare, China has to remain engaged in both economic and political workings of the Asia–Pacific region. Thus the Chinese leaders will learn the importance of continuous constructive engagement with regional peace and security. Economic slowdown may actually help their learning process.

NOTES

1 'More Achievements Expected in '94', *Beijing Review*, 10 January 1994, p. 4.
2 'The International Situation in 1993', *Beijing Review*, 3 January 1994, p. 19.
3 Wang Xin, 'Manzhouli: Future Metropolis on Northern Border', *Beijing Review*, 27 December 1993, pp. 12–16.
4 'Chinese Influx', *Far Eastern Economic Review*, 22 July 1993, p. 14.
5 Shigeo Hiramatsu, 'Kore ga "kaiyo haken" chizu da' (This is the map for 'maritime hegemony'!), *Shokun!*, July 1992, pp. 130–31.
6 Peter Hannam, 'Mongolia Fears China's Economic Clout', *FEER*, 17 June 1993, p. 52: and interview with a Mongolian security specialist, Tokyo, 13 June 1994.

7 'In Search of Genghis Khan', *FEER*, 2 September 1993, pp. 30–31.
8 Tsedendambyn Batbayar, 'Mongolia in 1993: Fragile Democracy', *Asian Survey*, January 1994, p. 45.
9 *Yomiuri Shinbun*, 25 June 1994.
10 'Sino-ROK Trade Balloons Sharply', *Beijing Review*, 28 March 1994, p. 28.
11 Nozomi Akizuki, 'Kai shisutemu no enchosenjo ni aru Chugoku Chosen Hanto kankei' (Chinese–Korean Peninsula relations on an extended line of the Chinese alien system), *Ajia Kenkyu*, 9:3, 1993, pp. 1–33.
12 *Sankei Shinbun*, 20 April 1993.
13 *Asahi Shinbun*, 11 June 1994.
14 *Sankei Shinbun*, 5 November 1992.
15 Michael Vatikiotis *et al.*, 'Gunboat Diplomacy', *FEER*, 16 June 1994, p. 23.
16 *Asahi Shinbun*, 9 July 1993.
17 *Sankei Shinbun*, 28 June 1994.
18 Bryce Harland, 'For a Strong China', *Foreign Policy*, no. 94, Spring 1994, pp. 48–52.
19 David Shambaugh, 'Growing Strong: China's Challenge to Asian Security', *Survival*, 36:2, Summer 1994, pp. 43–59.
20 Gerald Segal, 'China's Changing Shape', *Foreign Affairs*, 73:3, May/June 1994, pp. 43–58.

9 Chinese economic reform

The impact on policy in the South China Sea*

Michael Leifer

A positive link between Chinese economic reform and security policy has been demonstrated by the augmented capability of the People's Republic for projecting power in and through the South China Sea in pursuit of territorial and maritime claims. Success in economic reform has enhanced the ability of Beijing's national exchequer to disburse funds in support of this role by the armed forces in general and the navy in particular. Moreover, the South China Sea holds an intrinsic attraction for the government in Beijing because of the prospect of discovering and exploiting valuable natural resources, especially oil and gas, which would in turn make a major contribution to continuing economic development.

Those two aspects of linkage complement one another in that the benefits of economic reform permit greater support for defence expenditure and security policy which can be justified in part on grounds of future economic advantage. The fact that China became a net importer of oil during 1994 has reinforced such justification. Mixed opinions have been expressed about the potential reserves of oil in the area of the Spratly Archipelago.[1] These opinions would not seem to be shared in Beijing where a member of staff of the Foundation for International and Strategic Studies has drawn attention to an item published in May 1989 in the *China Geology Newspaper* which stated that surveys by the Ministry of Geology and Mineral Resources found that oil and natural deposits in the Nansha (Spratly) region amounted to about 17.7 billion tons. It was pointed out that 'This compares with 13 billion tons for Kuwait and ranks fourth in the whole world. No wonder the region is attractive'.[2]

It has been suggested that an additional link between economic reform and security policy in the case of the South China Sea has been the need to accommodate a disaffected domestic constituency. China's government has been said to cope with internal concerns over the impact of its economic reforms by taking a tough line on nationalist issues; 'hence Beijing's active and vigorous pursuit of claims in the South China Sea'.[3] However valid that suggestion, it is possible to question whether China's policy has been truly active and vigorous in military terms beyond gradually underpinning a tough declaratory position on maritime jurisdiction with an enhanced capability for power projection.

Irrespective of motive, economic reform has undoubtedly accelerated China's military modernization particularly since the end of the 1980s and to that extent facilitated security policy in the South China Sea. But the priority of sustaining the momentum of reform has also introduced a negative linkage and constraint on the robustness with which security policy is pursued. The South China Sea is subject to contested jurisdiction, especially the islands of the Spratly Archipelago which have attracted six claimants, if one includes Taiwan. Moreover, with the exception of Brunei which has not taken possession of any disputed territory, all claimant states have occupied islands of a kind and have underpinned that occupation with a military presence. Chinese economic reform, however, has been predicated on international cooperation, involving inward investment as well as trading ties, with states in Southeast Asia. At issue is the extent to which security policy expressed with reference to territorial and maritime ambitions in the South China Sea is fully consistent with the imperatives of economic reform in so far as those ambitions bring China into direct conflict with regional states. Should China succeed in realizing the full extent of its claims to sovereignty then it would be able to extend its jurisdiction some one thousand nautical miles from its mainland so as to command the virtual Mediterranean or maritime heart of Southeast Asia with far-reaching consequences for the strategic environment. That prospect is disturbing not only for those states which are in dispute with China over territorial and maritime jurisdiction. For example, in May 1994, Singapore's Senior Minister and former Prime Minister, Lee Kuan Yew, in an uncharacteristic admonition, publicly advised China to take steps to reduce regional anxieties over its intentions in the South China Sea.

China's dilemma in seeking to reconcile the priorities of economic reform with those of security policy has been eased somewhat by the evident difficulty confronted by regional states in concerting their thinking, let alone their policy, in facing up to the prospect of such an alarming projection of power and jurisdiction. Moreover, the piecemeal and relatively limited manner in which China has, so far, extended its jurisdiction southwards has added to the difficulty of mobilizing a countervailing regional response. The dilemma remains, nonetheless, an important latent factor bearing on China's security policy. For the time being, however, that dilemma has not become acute because China does not yet have the military capacity to mount a military seizure of the entire Spratly Archipelago.

CHINA AND THE SOUTH CHINA SEA

The extent of China's claims to territorial and maritime jurisdiction within the South China Sea has been articulated in some detail most recently in its 'Law on the Territorial Waters and their Contiguous Areas' which was adopted on 25 February 1992 at a meeting of the Standing Committee of the National People's Congress. That law made explicit that 'the extent of the PRC's territorial waters measures 12 nautical miles from the datum-line of the territorial waters' which is the breadth of territorial waters adopted in the United Nations Convention on the Law of the Sea which China has, so far, failed to ratify.

If the Pratas Islands and the submerged Macclesfield Bank (both contested with Taiwan) are set aside, two main sets of islands are at issue. The more northerly Paracel Islands have been subject to China's exclusive control since January 1974, albeit contested still by Vietnam and Taiwan. Only a limited part of the Spratly Archipelago is subject to the control of the government in Beijing and is the most important part of its *terra irredenta* in the South China Sea. This chapter will address in particular the issue of China's security policy in respect of the Spratly Islands whose full investment poses a major military problem because of their considerable distance from land-based points of military deployment and supply.

China's naval deployment within the Spratly Archipelago dates only from the early 1980s with an armed presence established on six islands towards the end of the decade.[4] But the claim to exclusive jurisdiction over the islands of the South China Sea dates from shortly after the establishment of the People's Republic. A formal statement was made by Premier Zhou Enlai in August 1951 in a hostile response to the publication of the draft Japanese Peace Treaty in which Tokyo's sovereignty over both the Paracel and Spratly Islands was renounced but without specifying their new ownership. Japan had occupied the Spratly Islands in 1938 on the grounds that they were Chinese territory, while at the end of the Pacific War the surrender of Japanese forces based there was taken by those of the Republic of China.

The position of the People's Republic has been maintained and reiterated with a steely consistency ever since Zhou Enlai's intervention. The idiom of claim to sovereignty has registered an adamant quality in which the adjectives indisputable and irrefutable have been used interchangeably. Such an adamant stand is relevant also to upholding a corresponding claim to the Senkaku or Diaoyutai Islands in the East China Sea. When circumstances have permitted, China has been willing to engage in a clash of arms to advance its territorial and maritime interests in the South China Sea.[5] Military force was employed first in January 1974 to consolidate control over the Paracel Islands well before China embarked on its policy of economic modernization which has generated additional funds for defence purposes in recent years. These northerly islands have the potential to serve as stepping stones to the Spratlys some 350 miles to the south. In August 1993, a Japanese newspaper published a satellite photograph showing what was claimed to be a Chinese airfield on Woody Island in the Paracel group.[6]

Economic reform has facilitated a more active security policy in the case of the South China Sea but the underlying irredentist goals were set by a longstanding view of the historical maritime domain of China inherited by the People's Republic from the Republic of China. Indeed, the positions of the governments in Beijing and Taipei on the South China Sea are identical. The continuous military occupation since 1956 of the sole island of Taiping or Itu Aba in the Spratly group by forces from Taipei has been tolerated in Beijing as if it were an act of trust undertaken in the overall Chinese national interest.[7] There was no protest from Beijing when the government in Taipei established a base for marines on Itu Aba in April 1994.

China's claim to the Spratly Islands is based on the historical grounds of discovery and administration. The act of discovery is said to go back to the Han dynasty in the second century with actual administration undertaken from the eighth century during the Tang dynasty. Although reference is made to naval expeditions in the thirteenth and fifteenth centuries, the effective Chinese presence probably amounted to little more than the activity of fishermen from Hainan Island, while French colonial intervention did not displace any administrative activity. A continuous administrative presence as required under international law cannot be demonstrated by China but the other claimants find themselves in the same difficulty. Vietnam, which also claims the entire Spratly group, seeks to inherit a tenuous French connection dating from a formal annexation in only 1933 which provoked a Chinese protest at the time. It has sought to bolster this basis for title through archeological excavations which purport to suggest a Vietnamese presence since the fifteenth century. China has maintained in addition, however, that a line of maritime demarcation in a treaty with France in 1887 placed the

Spratly Islands within its sovereign jurisdiction. Moreover, the government in Hanoi could find itself 'estopped' in any submission to an international tribunal. The Democratic Republic of Vietnam, which assumed the government of the whole of Vietnam in 1975 and changed in nomenclature to the Socialist Republic of Vietnam in 1976, officially acknowledged China's sovereign jurisdiction over the Paracel and Spratly Islands on two occasions during the 1950s. It only reversed its position after the unification of Vietnam in 1975, so arousing Beijing's anger at its alleged duplicity. Claims to parts of the Spratly Archipelago by the Philippines, Malaysia and Brunei are based primarily on unilateral interpretations of the changing law of the sea and lack of China's historical basis, however fragile.

Apart from a conviction that it possesses the only authentic claim, the government in Beijing appears fully conscious of the maxim that possession is nine-tenths of the law. Indeed, a factor which has driven the pace of security policy has been China's late entry in the contest for the Spratly Islands. All the other claimants, including Taiwan but with the exception of Brunei, were in possession of a number of islands, shoals, or reefs before China began to assert a measure of territorial control from the late 1980s. Moreover, that possession was secured by unilateral action involving the deployment and display of military force, most recently in February 1995 when the unoccupied Mischief Reef was seized in that part of the Spratly group claimed by the Philippines.

A resentment of double standards being applied to China's pursuit of a legitimate claim has been indicated in comment on regional responses to the promulgation of its maritime law in 1992 which implicitly empowered the use of force in upholding it. Lui Ning, a visiting research fellow at the Institute of Southeast Asian Studies in Singapore, has posed the question 'whether there was any country in the world whose maritime legislation did not contain a clause that authorized its armed forces to defend what is perceived as territorial waters?' More pointedly, he asked 'A number of recent Chinese actions have apparently caused more than a little discomfort in the region. But isn't a free-for-all grab-what-you-can in the South China Sea, the name of the game?'[8]

China's security policy has been driven by the requirement to get into the game before the onus for engaging in forcible possession at the expense of other claimants was placed exclusively on China. When China occupied and then garrisoned six islands during 1988, twenty-one other islands, shoals, and reefs had already been invested with a military presence by the Philippines, Malaysia and Vietnam in addition to the sole island initially invested by the government in Taipei in 1946.

Moreover, Chinese sources claim that Vietnam occupied another ten islands during 1988 and then five more islands between 1989 and 1991. The view from Beijing is that China, which has the only authentic title, has been denied its rightful territorial inheritance by the prior military assertiveness of other coastal states.[9]

THE ECONOMIC DIMENSION OF DEFENCE POLICY

It has been pointed out that 'The PLAN [PLA Navy] is embarking on a massive modernization programme and transition to a blue water power. Its objective is to become a new world-class Pacific power in the twenty-first century'. In this connection, it has been asserted also that 'Long before the navy achieves its long-term objective, the shift of its maritime strategy from brown water defence to green water defence has already exerted [sic] an impact on the regional power balance. The countries bearing the first brunt will be Vietnam and some of the ASEAN states which have territorial disputes with China over the Xisha (Paracel) and Nansha (Spratly) islands'.[10]

The build-up of blue-water capabilities, although enhanced of late, is not a new phenomenon originating from the fruits of economic reform. A concern to pre-empt Vietnam in the Paracel Islands before the unification of the country under a communist government as well as the need to cope with the Soviet presence in Cam Ranh Bay from 1979 were important considerations well before the benefits of economic reform became available for the armed forces. Indeed, a programme of military modernization specifically to enhance seaborne logistics capability relevant to upholding claims in the South China Sea was approved in 1977 before a formal commitment was made to economic reform. That programme has been undertaken and sustained on a systematic basis, if within budgetary constraints, and has been driven by a consistent territorial imperative.[11]

Obviously, greater budgetary largesse has enabled a more conspicuous defence build-up but this has facilitated the pursuit of a long-established policy rather than having determined it. A notable feature of the recent phase of a continuing build-up, however, is that it has proceeded in a strategic environment in which China has been freed of significant external threat to and constraint on its ambitions in the South China Sea. For example, a rusting vestige of the fleet of the former Soviet Union is all that remains in Cam Ranh Bay, while the United States withdrew its military presence completely from the Philippines before the end of 1992 and has refused to declare its intentions in the South China Sea. In that new environment, security policy has an added economic dimension in that greater budgetary provision will enhance China's facility for power projection in the South China Sea, while access to anticipated sources of oil and natural gas provides a strong and acknowledged motive and incentive for such enhancement.

It has been suggested above that the economic dimension of security policy in relation to the South China Sea has a direct bearing on its motivation and efficacy. Economic reform, however, has been based on full engagement with the region in attracting investment and securing markets. Moreover, with the end of the Cambodian conflict as a regional issue, China ceased to serve as a tacit alliance partner for the Association of South East Asian Nations (ASEAN) states and has been obliged to give more consideration to the consequences of alienating them over the South China Sea. At issue is the extent to which that risk might be taken, especially by a military establishment which also engages heavily in foreign economic relations. It is difficult to answer that question with any degree of confidence because of the likelihood of differing bureaucratic interests within a China without a powerful centre coming into contention over the matter.[12] For the time being, however, while not conceding any point of principle over sovereignty, China's government appears to have given some thought to the consequences of an unrelenting pursuit of its irredentist goals with economic interests presumably in mind. The occupation of Mischief Reef, however, marked the most southerly projection of a Chinese presence.

MANAGING THE REGION

China's assertiveness in the Spratly Archipelago became a matter of acute regional concern from March 1988 when a brief armed clash took place with Vietnamese forces as a result of an unplanned military encounter on Johnson Reef. There have been unconfirmed reports of a minor naval clash near Fiery Cross Shoal in the following December. China's investment had begun through placing observation towers on Fiery Cross Shoal in February 1988. By the end of the year, the People's Republic was in occupation of six islands, shoals, or reefs with markers placed on three others by mid-1992. Direct confrontation has occurred only with Vietnam, however, which responded by incorporating the Spratly Archipelago into Khanh Hao Province and occupying more islands. China may well have been constrained for a time after June 1989 by a concern with its international standing in the wake of the massacre in Tiananmen Square. But in its creeping assertiveness in the South China Sea, Chinese forces have not, so far, contested possession of any island with an ASEAN state; nor have they clashed again physically over territory with the Vietnamese. Moreover, China appears to have adopted a more emollient attitude towards the ASEAN states, including those with which it is in dispute over the Spratlys, by contrast with Vietnam. It has never failed to register a protest, however, when its claim to sovereign jurisdiction has appeared to be challenged. For example, a strong reaffirmation of sovereignty by Beijing followed the decision in May 1994 by the Philippines Department of Energy to grant an oil exploration permit in waters west of the island of Palawan to Vaalco Energy of the United States and its Philippine subsidiary, Alcorn Petroleum and Minerals.[13]

Vietnam has been directly challenged at the fringes of the Spratly Islands by the exploration concession granted in May 1992 to the Crestone Energy Corporation of Denver over which there appeared to be an attempt to exploit a residual American–Vietnamese alienation. That unilateral initiative appeared to contradict an earlier declared preference by China for setting aside the issue of sovereignty in favour of joint development with other claimant states. The extensive Wananbei-21 (or Vanguard Bank) concession is located in part of what the Vietnamese now claim is their continental shelf and exclusive economic zone in the Tu Chinh Reef area. The Chinese maintain, however, that the concession falls within the maritime domain of the Spratly Archipelago. Chen Bingqian, the spokesman for the China Offshore Petroleum Corporation has maintained that 'The area covered in the (Crestone) contract is within the territorial waters of the Nansha Islands which are within the jurisdiction of China and are within China's sovereignty', as well as reiterating somewhat incongruously his government's continuing interest in joint development.[14] By way of riposte in June and July 1992, Vietnamese forces made further territorial encroachments on two more reefs. China's commitment to Crestone has been sustained and in April 1994 the company announced that it had begun active exploration for oil in the disputed area 'with full support and protection from China'.[15] That announcement was made only hours before Vietnam signed contracts with a consortium of American and Japanese companies led by Mobil Corporation of the United States for exploring an adjoining westerly field within the area of China's maritime claim. Again, the Chinese have taken the position that they have been latecomers to oil exploration in the South China Sea which is their prerogative. It has been maintained that the Crestone contract in May 1992 is the first of its kind and has been entered into some decades after other countries have taken such initiatives. In May 1994, a Chinese Foreign Ministry spokesman denounced as illegal and as an infringement of sovereignty the oil exploration contract entered into between Mobil Corporation and Vietnam for the Blue Dragon field to the west of the Crestone concession. It is of interest to note that Mobil were drilling in this concession area during the early 1970s under authority from the government of South Vietnam without then attracting a protest from Beijing.

The validity of the Crestone concession would be a matter of dispute even if China's claim to the Spratly Islands were conceded. First at issue is that article in the Law of the Sea Convention of 1982 which maintains that 'Rocks which cannot sustain human habitation or economic life shall have no exclusive economic zone or continental shelf'. Second, Chen Bingqian as spokesman for the China Offshore Petroleum Corporation has implied above that his government's right to award the Crestone contract was because the concession area falls within the territorial waters of the Spratly Islands. However, China's territorial waters extend for only twelve miles while the distance between the nearest island in the Spratly group and the Crestone concession is well beyond twelve miles. It has been suggested that 'the Crestone award is symbolic of China's claim to the whole (South China) sea, which in effect would turn it into inland Chinese waters'.[16] Such a suggestion assumes a legal settlement to the contested jurisdiction which would seem highly unlikely. More relevant is the view by the same author that the Crestone contract is an attempt to set an eastward limit to Vietnam's oil search within its own continental shelf.[17] That view would seem to have been borne out by China's deployment of two warships in July 1994 to block the resupply of a Vietnamese oil-drilling rig in order to prevent it from working in the Wananbei-21 area.

Despite its measure of assertiveness, China had also indicated a disposition to conciliation over the South China Sea by sending unofficial representatives to confidence-building workshops inaugurated by Indonesia from mid-1990. But these annual occasions have proved frustrating for the government in Jakarta which has not been able to move the participants beyond well-meaning but ineffectual declaratory statements and an engagement in discussions of technical proposals.[18] It seemed initially as if China was engaged consciously in a divide and rule exercise between Vietnam and ASEAN, with landings on three more unoccupied islands occurring during the first half of 1992. Circumstances changed somewhat in July 1992 when Vietnam attended the annual meeting of the Association's foreign ministers in Manila. On that occasion, Vietnam, together with Laos, acceded to ASEAN's Treaty of Amity and Cooperation in South-East Asia which marked an important step towards full membership of the Association. Moreover, on July 22, its six foreign ministers issued an unprecedented Declaration on the South China Sea calling on claimant states to settle matters of contested jurisdiction by peaceful means. Although little more than a reiteration of a statement made at the Indonesian workshop held in Bandung in July 1991, it marked a significant diplomatic initiative which held out the prospect of political costs to any state which infringed its central axiom. Vietnam endorsed the Declaration with alacrity.

Although China was equivocal in its immediate response, and has never formally endorsed the Declaration, the tone of its approach to the South China Sea would seem to have softened subsequently without conceding an iota of sovereignty. China's own interests in the South China Sea had not changed but the need to manage the regional environment had become more pressing. The option of joint development, albeit without addressing the issue of sovereignty, was canvassed by Li Peng during a visit to Singapore in August 1990. Practical progress over a device employed with some success elsewhere by other coastal states has not been made and China has acted unilaterally over Wananbei-21 but the option continues to be mooted in Beijing.[19] As a participant in a growing network of regional cooperation, China's leaders have also found it politic to make explicit statements about foregoing the use of force in resolving the matter of jurisdiction over the Spratly Islands.

Such a position has been taken in negotiations with Vietnam even though there has been a failure to make tangible progress over their differences. The normalization of Sino-Vietnamese relations in November 1991 did not reduce bilateral tensions which were aggravated with China's assertiveness in the South China Sea in the first half of 1992. An attempt to improve the tone of the relationship was undertaken with a visit to Hanoi by Li Peng in December 1992 accompanied by recurrent commitments never to seek hegemony, an offer to engage in joint exploration and an agreement not to use force in resolving disputes. That initiative was succeeded in February 1993 by official talks on devising fundamental principles to be employed in such a resolution which culminated in an agreement on them in October that year. The second anniversary of normalization was marked by a visit to Beijing by President Le Duc Anh.[20] The net effect of such bilateral dialogue was Beijing's explicit acceptance of constraint in prosecuting a longstanding goal of security policy. More significantly for regional relations, a commitment ruling out the use of force in pursuing China's claim to the Spratly Islands was made in Kuala Lumpur in May 1993 by Minister of Defence Chi Haotian who was visiting with a delegation of senior officers, including Lt-General Li Xilin, the Guangzhou military area commander. He also remarked on the modest scale of

China's military budget and denied that the People's Republic intended to purchase aircraft carriers from the Ukraine and Russia.[21]

In that vein, China also responded positively to an invitation to participate in regional multilateral security dialogue. Its foreign minister took part in the inaugural meeting of the ASEAN Regional Forum in Singapore in July 1993 intended in principle to promote a structure of common security through the nexus of economic advantage joining all of East Asia. Such engagement, together with participation in the informal summit called by President Clinton in Seattle in November 1993 after the Asia–Pacific Economic Co-operation (APEC) ministerial meeting, implied accepting a measure of constraint on freedom of action in security policy. Through its commitment to economic reform, China was being drawn into an embryonic structure of good citizenship for the region whose implicit code of conduct could be violated only at some political and possibly economic cost. In July 1994, China's Foreign Minister, Qian Qichen, attended the first working session of the ASEAN Regional Forum in Bangkok where he stated that China stood for a gradual approach in finding practical means of regional security cooperation.

At issue are the interests which drive China in prosecuting security policy in the South China Sea in the light of the embryonic structure that the ASEAN Regional Forum represents. The specific interest of the navy in asserting and justifying a mission in a post-Cold-War world has been well identified and the ability of the government in Beijing to keep a firm grip on regional naval initiatives is uncertain. The extent to which China's navy through its regional command had begun to endanger the traditional right of innocent passage in its southern waters through acts ostensibly designed to counter smuggling became a minor international scandal in 1993. Such activity has been the subject of scrutiny by the International Maritime Bureau.[22] Reports of piracy in the South China Sea declined dramatically in the second half of 1993 but the practice which has not ceased altogether runs counter to Chinese protestations about not seeking hegemony. An additional special interest of note is that of the China Offshore Oil Corporation which has joined in economic partnership with Crestone. It has done so with the apparent full endorsement of the government in Beijing which is the locus of authority in security policy.

For the time being, China has limited its assertiveness in the Spratly Archipelago. It has established its stake there in order to ensure that no attempt to settle the dispute over jurisdiction can take place without its participation. China has also laid down a marker to Vietnam in particular over the extent of oil exploration it will tolerate. Moreover,

China has made clear its preference for bilateral as opposed to multilateral negotiations in order to avoid placing its representatives in a position of diplomatic disadvantage. Military modernization continues with an evolving capability for engaging in power projection, although its momentum may be affected by the recent difficulties encountered by the economy. China's navy has, of course, enjoyed a free rein within the South China Sea from the early 1980s when it first conducted an exploratory expedition in the interest of asserting sovereignty as far south as James Shoal off the Borneo coast. That kind of exercise in naval display has been repeated subsequently more than once. China's ability to engage in sustained military operations in the face of effective resistance has yet to be demonstrated, however.

The critical problem long identified is that of superiority of air power and cover arising from the limited range of shore-based aircraft. Apparent interest in both aircraft carriers and short take-off vertical landing aircraft to cope with that problem has been noted as 'alarming military news to China's neighbours'.[23] Recent commissions of naval vessels and procurements of aircraft from Russia which would go some way to meeting that problem have been set out in *The Military Balance*, which addresses speculation about China's interest in an aircraft carrier and cautions 'it will be several years before China could have an operational aircraft carrier and air wing'.[24] Given current limitations in the sustained projection of military power, China's option of engaging in creeping assertiveness according to military capability and opportune circumstances would seem constrained also by the requirement to make a judicious assessment of the importance of conciliating regional neighbours in the interest of economic reform. An additional dilemma for China is that the realization of military modernization may come too late to deny exploitation of natural resources to regional states, some of whom have been able to afford to procure state-of-the-art weapons with which to defend their maritime space. In the meantime, it is quite clear that China is prepared to risk disconcerting some regional neighbours as it singles out Vietnam for special harsh treatment in the interest of protecting its offshore oil concessions.

REGIONAL RESPONSES

China's ability to pursue its interests in the South China Sea has been enhanced by the change in the regional strategic environment after the end of the Cold War. Although the military clash with Vietnam in March 1988 would not seem to have been planned, the reluctance of the Soviet Union to honour its treaty obligations was conspicuous. Correspondingly, the United States under George Bush and Bill Clinton has failed to indicate any enthusiasm for assuming a military balance of power role in the South China Sea. Indeed, Secretary of State James Baker was present in Manila in July 1992 for the ASEAN Post-Ministerial Meeting but did not seem to wish to be drawn into passing judgement on the merits of the prior declaration on the South China Sea in case it might imply a security commitment. His successor Warren Christopher has not been any more forthcoming on American responsibilities in the South China Sea.

In the absence of suitable access to external countervailing power, regional states have been thrown back on their own resources in an attempt to contain China from asserting its way into the maritime heart of Southeast Asia. The problem for ASEAN in concerting a response is not one which has been precipitated by the contest for the South China Sea. The ASEAN states have long been afflicted by an inability to assume a common position over China which was exemplified by intramural tensions during the Cambodian conflict. The position has been expressed well by Tim Huxley who has pointed out:

> Beneath the superficiality of a common ASEAN position on the South China Sea, opinion within ASEAN regarding this complex issue is quite diverse. Apart from the problem that none of the three ASEAN members with claims in the Spratlys show any inclination to compromise on the issue of sovereignty with each other, ASEAN's membership remains divided in its attitude towards China.[25]

In addition to the regional division in outlook towards China, there is also the important underlying problem of rivalries among the ASEAN states which inhibits security cooperation.[26]

The absence of a common strategic perspective among the members of ASEAN has long been a fundamental obstacle to any serious consideration of alliance practice. Practical security cooperation among ASEAN states takes place primarily on a bilateral basis but never under the aegis of the Association. There has been an evident enhancement of national defence capabilities on the part of all member states, with the exception of the Philippines, but not as the result of joint initiatives with the prospect of combined operations against an extra-mural adversary.[27] In the absence of a common will to address the problem of an assertive China through the mechanism of the balance of power, the ASEAN states collectively have been obliged to rely *faute de mieux* on the alternative stratagem of common security but without any indication of an ability to address the issue of arms control.

Over regional security, Gerald Segal has enquired 'The wider question for many Southeast Asians is whether China is better placated or deterred'? He goes on to answer his own question in asserting that 'without a balance of power, Southeast Asians are vulnerable', and supports his position by pointing out that 'arguments that increasing economic interdependence will keep China from using force are simply not borne out by recent events'.[28] The fact of the matter is that Southeast Asians do not have the facility to conjure up a balance of power through appropriate extra-regional affiliation. The United States is seen to have lost the will to uphold the regional balance of power, Russia does not count militarily in South-East Asia, while Japan is viewed as a sleeping giant best left alone. Southeast Asians also lack the common interest and purpose to contemplate such an undertaking on an intra-regional basis. For that reason, they are not in a position to address Gerald Segal's wider question. They are not in any position to engage in deterrence, nor are they of one mind over placation. Deterrence arises only in so far as China's willingness to violate the canons of common security could impose a diplomatic and economic cost.[29]

Despite his realist disposition, Gerald Segal approves of the 'desirability of encouraging a China that is peacefully interdependent with the world beyond its borders'.[30] This is the course of action which ASEAN has adopted as the collective stratagem for a post-Cold War regional order through its initiative for a Regional Forum with China's participation. That strategem is at an embryonic stage and it remains to be seen whether it will evolve into a viable structure of constraint which will become a serious factor in China's calculations. In the meantime, the more pragmatic of ASEAN's foreign ministries believe that China will not hesitate from flexing its military muscles if necessary, despite a continuing preoccupation with economic reform.

CHINA LOOKS SOUTH

It has been argued that China's security policy with reference to the South China Sea has its roots in a determination to consolidate the national domain and to secure economic advantage. That policy has been pursued through changing strategic circumstances and economic priorities and has been facilitated by the success of a reform programme which has released additional resources for military modernization and power projection. By contrast with the differing phases of the Cold War, however, China does not now encounter the same degree of military opposition in seeking to extend its territorial and maritime jurisdiction but has yet to overcome fundamental logistical obstacles to effective power projection. It has also to face up to the political and possible economic opportunity costs of an untramelled assertiveness.

Despite lacking ideal means for power projection, one cannot rule out China becoming more assertive with the military capability which it currently possesses as a consequence of any internal political change. The inability of Southeast Asian states to adopt a common position over the South China Sea beyond a superficial declaratory stance could encourage such a course of action. Moreover, the effect of assertiveness by China could well be to sow the seeds of further regional disarray rather than to engender any common action given the tacit understanding that there is no disposition to confront China. That said, in the case of the Spratly Islands, apart from the armed clash with Vietnam in 1988 which did not seem to have been planned, China has, so far, engaged only in a relatively limited piecemeal assertiveness in circumstances where there has been minor risk of a military engagement. None of the ASEAN states have been confronted militarily. Moreover, Vietnam has not yet been directly challenged in an attempt to coerce it into relinquishing possession of any islands which it has invested although its oil exploration activity has been subjected to naval harassment in the area of its continental shelf.

Whether this current moderation in assertiveness is a consequence of a limitation in military capability or a product of an interest in regional cooperation inspired by the priority of economic reform or both remains to be seen. The fact of the matter is that China has never shown the slightest interest in compromise over sovereignty in the South China Sea, despite the hegemonic implications of its position. That lack of compromise has been held consistently irrespective of the mixed impact of economic reform on security policy. At issue is whether an embryonic architecture of regional security based on an economic nexus to which China has become a party will serve as sufficient constraint on a longstanding nationalist purpose sought through security policy? China's past record is not encouraging in this respect, particularly as its policy is driven by a strong sense of frustrated territorial entitlement. In addition, its military capability is being progressively enhanced within a regional environment in which the conventional countervailing forces of the balance of power are not actively at play.

NOTES

* Another version of this chapter has been published in *Survival*, vol. 37, no. 2 (Summer 1995), entitled 'Chinese Economic Reform and Security Policy: The South China Sea Connection'.

1 See, for example, the letter by E. F. Durkee, a petroleum geologist, *International Herald Tribune*, 14 April 1992; and William Mellor, 'Tug of War', *Asia, Inc*, Bangkok, September 1993, pp. 54–8.
2 See Pan Shiying, 'The Nansha Islands. A Chinese Point of View', *Window*, Hong Kong, 3 September 1993, p. 28.
3 See Gerald Segal, *China Changes Shape: Regionalism and Foreign Policy*, Adelphi Paper 287, IISS/Brassey's, London, 1994, p. 45.
4 The best recent account of China's naval assertiveness in the South China Sea is to be found in John W. Garver, 'China's Push Through the South China Sea: The Interaction of Bureaucratic and National Interest', *The China Quarterly*, December 1992.
5 For a comprehensive account of China's position, see Chi-kin Lo, *China's Policy Towards Territorial Disputes. The Case of the South China Sea Islands*, Routledge, London and New York, 1989; also Jeanette Greenfield, *China's Practice in the Law of the Sea*, Clarendon Press, Oxford, 1992, Chapter 7.
6 See the reference to *Yomiuri Shimbun, International Herald Tribune*, 5 August 1993.
7 For a Taiwanese view which indicates a prospective compatibility with the PRC, see Peter Kien-hong Yu, 'Issues on the South China Sea: A Case Study', *Chinese Yearbook of International Law and Affairs*, vol. 11, 1991–2.
8 See, *Business Times*, Singapore, 29 October 1992.
9 See Pan Shiying, op. cit.
10 See You Xi and You Xu 'In Search of Blue Water Power: The PLA Navy's Maritime Strategy in the 1990s', *The Pacific Review*, vol. 4, no. 2, 1991, pp. 147–8.
11 See Garver, op. cit.
12 See Garver, op. cit.
13 See *Far Eastern Economic Review*, 30 June 1994.
14 See BBC *Summary of World Broadcasts*, FE/2003 G/1–2.
15 See *International Herald Tribune*, 21 April 1994.
16 See Philip Bowring, 'China Is Getting Help In a Grab at the Sea', *International Herald Tribune*, 6 May 1994.
17 Bowring, op. cit.
18 See Mark J. Valencia, 'Spratly Solution Still at Sea', *The Pacific Review*, vol. 6, no. 2, 1993.
19 See Pan Shiying, op. cit., pp. 30–34.
20 For an succinct account of the more recent condition of Sino–Vietnmese relations with reference to the South China Sea, see Carlyle A. Thayer, 'Vietnam: Coping with China', *Southeast Asian Affairs 1994*, Institute of Southeast Asian Studies, Singapore, 1994.
21 See BBC *Summary of World Broadcasts*, FE/1699 A1/1.
22 See *Piracy Report*, (1 January–31 December 1993), International Maritime Bureau, Barking, UK, 1994; Barry Wain, 'China's Gunboat Diplomacy Must Be Stopped', *The Asian Wall Street Journal Weekly*, 28 March 1994; also *Far Eastern Economic Review*, 16 June 1994.
23 Jencks, op. cit., pp. 101–2.
24 *The Military Balance 1993–1994*, International Institute of Strategic Studies, London, 1993, p. 148.

25 See Tim Huxley, *Insecurity in the ASEAN Region*, Royal United Services Institute for Defence Studies, London, 1993, p. 34.
26 See Lee Kim Chew, 'Rivalries Among ASEAN States Inhibit Cooperation', *The Straits Times*, 8 April 1994.
27 See Huxley, op. cit., p. 66.
28 Segal, op. cit., p. 46.
29 See the argument in Amitav Acharya, *A New Regional Order in South-East Asia: ASEAN in the Post-Cold War Era*, Adelphi Paper 279, IISS/Brasseys, London, 1993, pp. 33–40.
30 Segal, op. cit., p. 58.

10 Arms races and threats across the Taiwan Strait

Michael D. Swaine

In a regional environment marked by the end of the Cold War, the expansion of democratic regimes, and growing diplomatic and economic cooperation among states, the Taiwan–China military confrontation arguably poses one of the very few tangible threats to regional stability and peace remaining in East Asia.[1] In fact, the possibility of military conflict across the Taiwan Strait has become an urgent concern in some quarters in recent years, largely as a result of rapid, and in many ways revolutionary, domestic changes occurring on both sides of the Strait. Economic reform in China has led to the highest growth rates in the world, greater Chinese self-confidence in the international arena and increased efforts at military modernization in areas that potentially threaten Taiwan's security. Equally significant, the notion of correcting past historical humiliations and defending or asserting Chinese claims over nearby territories is increasingly creeping into Chinese leadership pronouncements, causing greater anxieties in many Asian capitals, including Taipei.

On Taiwan, democratization from above has led to the Taiwanization of the political process and resulting demands for independence from the mainland, voiced most loudly by members of the increasingly popular Democratic Progressive Party (DPP). Moreover, continued high growth rates, expanding levels of foreign trade and investment across the region, and the accumulation of enormous foreign exchange reserves have given Taiwan new avenues for asserting its influence in the regional and global arena. Both of these recent trends are viewed with growing concern by Beijing, despite the phenomenal increase in cross-Strait contacts and economic activity that has taken place since the mid-1980s.

Finally, concerns are further heightened by signs of confusion in the US position toward Taiwan. Recent decisions by Washington to provide significant amounts of advanced weaponry to Taipei and to expand contacts with Taiwanese officials suggest a reversal of earlier US commitments to steadily reduce the level of military assistance provided to the island, and to strictly limit official US–Taiwan interactions. At the same time, many knowledgeable analysts believe that US incentives to come to Taiwan's aid in a conflict have diminished significantly in recent years. Such factors, when placed in the context of domestic developments in Taiwan, are seen by some to strengthen Beijing's willingness to use coercive diplomacy, if not outright force, to reunify the nation. The dangers of this situation were most recently confirmed by the crisis prompted by President Clinton's decision, under pressure from the US Congress, to grant Taiwan President Lee Teng-hui a visa to visit Cornell University in June 1995. This action significantly damaged US–China relations and prompted a threatening series of Chinese military displays toward Taiwan, thus greatly raising the level of military tensions across the Taiwan Strait.

This chapter examines the changing military and diplomatic relationship between China and Taiwan and assesses the potential for a further escalation of military tensions, perhaps leading to a full-blown arms race and/or armed conflict. Particular attention is given to the impact of developments within China upon the future threat environment, given the central importance of Beijing's capabilities and intentions to the overall calculus.

Before proceeding, it is essential to define an arms race and to identify the problems it poses. As Colin Gray has stated in his classic work on the subject, an arms race must satisfy four basic conditions:

1 There must be two or more parties who are conscious of their antagonism.
2 They must structure their armed forces with attention to the probable effectiveness of the forces in combat with, or as a deterrent to, the other arms race participants.
3 They must compete in terms of quantity (men, weapons) and/or quality (men, weapons, organization, doctrine, deployment).
4 There must be rapid increases in quantity and/or improvements in quality.[2]

Obviously, the *interactive* nature of military or military-related acquisitions (underlying condition two) is of particular importance in the definition of an arms race. However, this suggests the need for caution when considering the case of China and Taiwan. It is immediately apparent that the enormous difference in strategic outlook between the two antagonists makes it extremely difficult to identify with complete certainty the existence of any specifically interactive pattern of military escalation between the two sides. While Taiwan's military posture is almost entirely keyed to the threat from across the Strait, China's posture is intended to deal with an extremely wide range of threats from many countries, both nuclear and non-nuclear, distant and nearby, as well as to serve an array of other national security interests. Any examination of Beijing's military capabilities toward Taiwan must thus include overall developments in relevant areas such as air and naval power, since an attempt to identify only those forces likely to be used against the island is virtually impossible.

THE 1980s: AN INCIPIENT ARMS RACE?

During the 1980s, military developments in China and Taiwan did not even approximate an arms race in the above sense, nor did they constitute any other pattern of escalating military tension.[3] First, as Table 10.1 indicates, arms expenditure trends moved in opposite directions. Taiwan increased its level of spending almost continuously during the decade, at an average rate of over 4.5 per cent per year. This spending amounted to at least 6 per cent of annual GDP throughout the decade. In contrast, Chinese spending actually declined at an average annual rate of over 1.5 per cent during the same period, with yearly totals usually amounting to significantly less than 4 per cent of annual GDP.[4] Moreover, Taiwan imported over US$7.7 billion in arms during the decade, while China purchased less than half that amount.[5]

Table 10.1 Annual percentage change in military expenditure: China–Taiwan

	1981–2	*1982–3*	*1983–4*	*1984–5*	*1985–6*	*1986–7*	*1987–8*	*1988–9*	*1989–90*	*Decade average*
China	2.6	−1.3	−1.3	−5.3	−1.4	−4.2	1.5	−15.9	10.3	−1.6
Taiwan	12.8	0.9	−0.7	10.4	3.2	3.3	7.8	−1.0	4.5	4.57

Source: Based on figures found in *SIPRI Yearbook, 1991: World Armaments and Disarmament*, Stockholm International Peace Research Institute, Oxford University Press, pp. 157, 169–73.

This contrasting pattern of spending reflected two very different military doctrines and sets of strategic priorities. The abrogation of the US–Taiwan Mutual Security Treaty in 1980 (following the establishment of formal diplomatic relations between Washington and Beijing in 1979) forced Taipei to adopt a purely defensive military strategy toward Beijing, thus discarding any remaining pretence toward an offensive doctrine based on recovery of the mainland, with possible

US support. This new strategy was centred on acquiring the capability independently to defend Taiwan by achieving air superiority over the Strait and naval approaches to the island, and by applying air and naval power to break a possible blockade and/or prevent an amphibious landing.[6]

Such a strategy required major improvements in Taiwan's naval and air capabilities, (intended in large part to compensate for the loss of major military purchases from Washington[7] and to replace ageing equipment), and the creation of an independent defence industrial base. This effort centred on both foreign purchases from new sources in such areas as Europe, Israel, and Singapore, and stepped up efforts to develop self-reliance capabilities in a variety of weapons systems and technologies.

Taiwan's efforts in the former area often encountered serious difficulties during the decade, largely because of Chinese pressure on potential suppliers. However, major foreign purchases in the 1980s included two Dutch submarines, while foreign technologies were used to refit old destroyers and frigates. Taiwan's effort at self-reliance centred on a programme for the development of an indigenous defence fighter (IDF) with US technical assistance not prohibited by the 1982 Communiqué. The programme was inaugurated in 1982, when the Reagan administration had made an initial decision not to sell fighter aircraft to Taiwan.

In short, during the 1980s, Taipei made concerted attempts to close a perceived window of opportunity for Beijing to increase its military capabilities *vis-à-vis* Taiwan when the latter's diplomatic fortunes declined in the early 1980s. China, however, did not embark on a crash programme to take advantage of this opportunity. Throughout the 1980s, military modernization was accorded a very low priority in Beijing, as resources and energies were devoted to civilian economic reform and the establishment of expanded, peaceful relations across the Asia–Pacific and beyond. In addition, China's limited defence funds were used in support of a new and more complex military doctrine of 'local war' and 'active peripheral defence' intended to address a wide variety of possible contingencies, not just an attack against Taiwan.

China's new defence doctrine emerged gradually over the early and mid-1980s, largely as a result of Beijing's altered domestic priorities, the declining Soviet threat (which downgraded the chances for a global war), the emergence of a less certain strategic environment in Asia, and, more directly, the poor showing of Chinese troops during the Sino–Vietnamese border conflict. This doctrine supplanted the past, outmoded Maoist notion of People's War, which had relied heavily on the use of huge, and rather primitive, infantry-based land forces in a protracted contest of attrition against a full-scale invasion. It required the development of advanced weapons with medium- and long-range force projection, mobility, rapid reaction, and offshore manoeuvrability capabilities. As a result, the focus of Beijing's arms and technology acquisitions shifted from the continued improvement of the PLA's massive ground forces to the attainment of much smaller but highly proficient ground-based rapid reaction units, relatively sophisticated air and naval forces, and a modern combined service tactical operations doctrine. These new forces were intended to cope with a variety of contingencies, including internal social unrest, land-based border conflicts in Inner Asia and elsewhere, the defence of territorial claims in the South China Sea (and other offshore areas), and the uncertainties posed by the changing military posture of the USA, Russia, Japan, and perhaps India. Although 'the liberation of Taiwan' was also usually included in this list of security concerns it was not given a high priority. Instead, from at least the perspective of the PLA Navy (PLAN), operations in the South China Sea became the primary mission for China's future air and naval forces, once the Soviet naval threat had largely evaporated.[8]

In sum, Taiwan made some major gains in acquiring essential defensive systems during the 1980s, while China focused its energies primarily on non-military matters, or on military contingencies other than the Taiwan Strait, requiring gradual qualitative improvements. In fact, far from entering into an interactive pattern of force acquisitions, Taiwan and the mainland made a positive breakthrough of sorts in their relationship during the decade, leading to an explosion in economic, cultural, and person-to-person contacts. Beijing intensified its 'peace offensive' towards Taiwan, beginning with the well-known, nine-point proposal of 1981, intended to establish a 'one country, two systems' arrangement between the two sides. Taiwan rejected these overtures but proceeded to implement an increasingly flexible stance toward the mainland in a variety of areas, most notably trade and travel.[9] As a result, tensions between Beijing and Taipei eased significantly, and mutual incentives – and expectations – grew for a peaceful resolution of their differences.

THE 1990s: A TURN FOR THE WORSE?

Despite the continued growth of both unofficial and official contacts and trade between both sides, military developments in China and Taiwan began to move in an ominous direction in the late 1980s and early 1990s. The pattern of defence spending and the composition of force acquisitions on both sides of the Strait took on at least the appearance of an incipient arms race. These acquisitions eventually resulted in the current military balance of air and naval forces presented at the end of this chapter (Table 10.4).

Taiwan's pace of military expenditure accelerated in the early 1990s, with average increases of more than 10 per cent annually.[10] Moreover, Taipei reportedly plans to maintain this pace throughout the remainder of the decade, eventually spending a total of approximately US$40 billion on arms.[11] Of greater significance, Chinese military expenditures began to increase in a similar manner, thus reversing the sluggish trend of the 1980s. Increases have averaged well over 10 per cent since 1989, and reports suggest that the Chinese leadership intends to maintain such increases for at least the remainder of the 1990s, as in the Taiwan case.[12]

This rate of spending is intended to support a range of significant weapons system and technology acquisitions on both sides of the Strait. Taiwan has embarked on several major programmes to improve its naval and air capabilities that build significantly upon the initial gains of the 1980s.[13] The Kwang Hua I Programme includes the construction of at least 8 improved Perry-class (4,100 tonnes) frigates of US design, with the possibility of an additional 8 in the future. Each frigate has advanced anti-aircraft capabilities and is able to carry 2 anti-submarine warfare (ASW) helicopters. Production is scheduled at roughly one ship per year. Three had been commissioned by March 1995, and the eighth ship is due by the late 1990s. In the meantime, Taiwan is leasing 6 to 9 Knox-class (3,000 tonnes) ASW frigates from the US.

The Kwang Hua II Programme includes the assembly of up to 6 modern, French-designed, Lafayette-class (3,200 tonnes) frigates, with the option of acquiring a further 10. In addition, Taiwan is acquiring a significant number of modern US attack and ASW helicopters, is upgrading its 40 or so maritime reconnaissance and ASW naval aircraft and has ordered 4 coastal minehunters from Germany (reportedly disguised as commercial vessels)[14] to supplement its fleet of 22 mine countermeasure (MCM) ships. Furthermore, Taipei intends to purchase 6 to 10 attack submarines[15] to supplement its small fleet of 4, as well as 10 corvettes and 12 missile-equipped fast attack craft. Although already possessing a significant blue-water capability, such acquisitions will transform the Taiwan Navy into 'a modern, relatively sophisticated naval force' by the end of the decade.[16]

Taiwan's actual and planned air power acquisitions are almost as impressive. In 1992, the US altered its cautious arms sale policy of the

1980s and sold 150 F-16A/B fighter aircraft to Taiwan, intended to replace the island's forty-year-old F-104s and thirty-year-old F-5Es. The first group of F-16s will be delivered in October 1996 and the last by the end of the century.[17] This 'breakthrough' led to a French sale of 60 Mirage 2000–5 multirole aircraft soon thereafter, to be delivered between 1993 and 1998. Other foreign governments are now reconsidering their past cautious policies toward major military sales to Taipei. In addition to these purchases, Taipei is going forward with the IDF programme, although at a slightly slower pace and with fewer aircraft in mind. It has apparently discarded the original plan to manufacture 250 fighters, intending instead to produce about 130 to 140 by the end of the decade, adding to the 10 test versions already in operation. The above aircraft acquisitions, plus the retention of up to 100 F-5Es already in service, mean that 'Taiwan will have one of the most powerful and modern air forces in East Asia, with a total of around 450 aircraft'.[18] In addition, to improve its air defences, Taiwan has contracted with a US company to co-produce a Patriot missile derivative, replacing its outdated Nike system, and has deployed an indigenous surface-to-air missile capable of intercepting advanced fighters beyond visual range.[19]

Beyond these air and naval weapons acquisitions, Taiwan is also carrying out major improvements of its early warning, surveillance, and operational command and control capabilities. It will reportedly continue to phase out ageing C-119s while purchasing new C-130Hs, airborne warning and control systems (AWACs) and Early Warning (EW) aircraft. It also has plans to improve the automation of its C^3I systems.[20]

In China, the reversal of defence spending trends has led to much greater activity on several fronts. Most major acquisitions remain keyed, as in the 1980s, to the development of naval and air power projection capabilities as well as more sophisticated defensive armaments.[21]

Movement toward the attainment of a high seas naval capability picked up speed in the late 1980s and early 1990s, with the acquisition of a wide range of weapons systems and technologies. The PLAN has added nearly 20 principal surface combatants (i.e. ships with at least 1,000 tonnes displacement) to its inventory since the mid-1980s. It is acquiring a new class of destroyer (the Luhu or Type 052), is upgrading versions of its mainstay Luda-class destroyers, and is developing a new class of missile frigate (the Jiangwei). These vessels possess missile capabilities (including silkworms and surface-to-air missiles), anti-missile missiles, and more sophisticated radar and fire control and ASW capabilities, as well as electronic countermeasures.[22] The PLAN is also developing new classes of resupply amphibious assault ships and missile patrol craft and is greatly expanding its number of mine warfare ships.[23]

China's total number of operational conventional submarines has probably dropped by over one half, from one hundred in the mid-1980s to less than fifty today. The vast majority of the remaining subs are outdated Romeo-class models from the late 1950s. However, the PLAN is endeavouring to upgrade the quality of its submarine force, first by improving its Ming-class submarines,[24] and then eventually by developing a new type of diesel-electric sub to replace the Romeo and Ming classes. This indigenously-built Song-class sub includes advanced Russian, French and possibly Israeli technologies. The first hull was launched in 1994. China has also purchased 4 sophisticated Kilo-class conventional subs from Russia and might eventually acquire up to 18 more.[25] Improvements are also underway in China's nuclear submarine fleet. These include plans to supplement (or probably in some cases replace) the PLAN's 5 Han-class nuclear attack submarines, and to produce an additional Xia-class nuclear ballistic missile submarine with an improved missile. There have also been persistent reports that the Chinese plan to construct or purchase 2 medium to large (40–50,000 tonne) aircraft carriers, for deployment by 2010 to 2020.[26]

These acquisitions have brought about major improvements in several areas. The operational range, fire power, and air defence capability of principal surface combatants has increased considerably, *theoretically* permitting many destroyers and frigates to operate with minimal air cover.[27] Moreover, as a result of these and other improvements, the PLAN has extended its capability to carry out more sophisticated operations farther from shore and for longer periods. For example, the PLAN has conducted multi-ship task force operations and fleet exercises in recent years, involving surface, subsurface, and aviation assets. Operations included maintaining and breaking blockades, attacking pipelines, shooting at mines, etc.[28]

In contrast to the above successes on the naval front, progress in improving China's air capabilities has been highly limited, although future plans remain ambitious. Beijing continues to rely primarily upon obsolescent versions of the Soviet MiG-17, MiG-19, and MiG-21 fighters and of the Soviet TU-16 bomber. These are known, respectively, as the J-5, J-6, J-7 and H-6. (The vast majority (about 3,000) of China's fighters are J-6s.) The PLA Air Force (PLAAF) also operates a small number (approximately 100) of the more advanced J-8 fighters, as well as about 500 low-performance Q-5 ground attack aircraft. China is reportedly working to upgrade the H-6 bomber into a multirole interceptor naval strike aircraft capable of sending an air-launched cruise missile. But even if successful, output is expected to be extremely low.[29] Production rates for the workhorses of the PLAAF fighter force are also very low, and will likely remain so: about 40 J-7s and 12 J-8s per year (production of the J-6 apparently ceased in the late 1980s). Moreover, the J-8 (roughly comparable to a basic F-4) continues to be plagued by engine and fuel consumption problems and a poor weapons system. Overall, Beijing's efforts to develop an advanced indigenous fighter and combat aircraft industry have been largely unsuccessful, and there are few signs of a breakthrough in at least the near future.[30]

At least partly to compensate for its air power problems, China has purchased 26 Su-27 aircraft from Russia and may obtain a further 48 by the end of the decade. Beijing is also reportedly arranging with Russia to manufacture under licence an estimated 300 additional Su-27 aircraft.[31] The Su-27 is an all-weather, counter-air fighter capable of operating from a carrier, with ramp assisted takeoff. It has an integrated fire control system, look-down, shoot-down radar, and is refuellable in flight. The Chinese also intend to co-produce a hybrid version of the Israeli Lavi fighter, termed the F-10, possibly with both Israeli and Russian assistance.[32] This multi-role fighter-bomber is almost identical to the F-16 in its interceptor and ground attack roles. Successful incorporation of these two advanced fighters would significantly improve Beijing's air combat capabilities.[33]

Finally, the PLAAF and PLAN are making concerted efforts to develop or purchase various 'force multipliers', especially airborne early warning and mid-air refuelling capabilities. For example, a British company is reportedly outfitting the H-6 bomber with an air-to-air refuelling system, to be used with the Q-5 ground attack aircraft. Other components of such a technology are also reportedly being supplied by Israel. Some specialists have estimated that the PLA will attain an air refuelling capability for about two squadrons (24–30) of aircraft (probably J-8s) within five years.[34] This will considerably expand the range of at least a small portion of China's fighter force beyond its current general operational distance of about 250–300 miles.

Several of these improvements in advanced naval and air power and amphibious lift capabilities are related to China's ongoing effort to develop an array of rapid reaction units (RRUs), as part of the doctrine of 'local war' and 'active peripheral defence' outlined above. These RRUs include both several battalion-sized, infantry-based units assigned to various Chinese group armies, and a special marine force attached to the PLAN, supported by amphibious assault ships and landing craft.[35] At least one well-informed expert believes that Beijing has the lift and transport capability to project about two division equivalents (about 25,000 troops) 'a good distance' from its shores, and can probably conduct an amphibious landing of *nearly* division strength (i.e. 10,000–14,000 men) 'well away from its immediate territorial waters'.[36] Most specialists on the Chinese military question such estimates, however, and insist that, although having made considerable progress in recent years, the PLA is far from attaining the capability to project two divisions.

In addition to these improved infantry-based power projection capabilities, Beijing has also been improving its ability to conduct coordinated operations among several services, including those in support of amphibious landings. Since 1993 the PLA has conducted a series of increasingly sophisticated and extensive military exercises, many along China's eastern and southern coastlines. These exercises involve ground forces, marines, airborne drops in support of amphibious operations, and air forces. Amphibious landing exercises have reportedly involved a full regiment, rather than a single battalion, as in the pre-1993 period. This suggests that China might attain a full divisional amphibious landing capability within two to three years.[37]

Finally, beyond the above actual and planned improvements in naval and air forces, China has also developed a new class of conventionally armed, mobile, tactical surface-to-surface ballistic missiles with ranges of around 180–350 miles. These place Taiwan (and other nearby areas) within range of a missile attack from the mainland.[38]

The above military acquisitions on both sides of the Strait certainly suggest that China–Taiwan military behaviour is taking on more of the attributes of an arms race. The pace of acquisitions has increased, and the actual composition of the weapons systems being acquired suggests that a more interactive process may be at work in some instances. As the above indicates, the PLA is steadily modernizing and improving its operational capabilities in several critical areas required to launch an attack against Taiwan. These include blue water surface operations necessary for both invasion and blockade, the development and use of stand-off and precision-guided, air-to-surface and surface-to-surface munitions, combined force tactics, including amphibious landings, and improved command and control facilities. At the same time, Taiwan is acquiring a new generation of advanced aircraft, naval vessels, missiles, radars, ASW capabilities, command and control centres, and other weapons systems that will be operational toward the end of the 1990s.[39]

Statements or concerns expressed on both sides also seem to support the notion that such acquisitions might amount to the beginnings of an arms race. The recent White Paper issued by Taiwan's Ministry of Defence clearly indicates that Taiwan believes it is facing an increased threat from Beijing as a result of the PLA's recently inaugurated modernization efforts. The document cites, among other examples, the increased possibility of an airstrike, a missile attack, or an attempted naval blockade of Taiwan.[40] Moreover, high-level Taiwan military officers and other concerned observers insist that the PLA's recent acquisitions have virtually eliminated Taiwan's past qualitative edge in many weapons systems and that a new window of opportunity has opened for the mainland to apply superior force to the island.[41] Some observers often add that Beijing now has more forces to draw upon for service in the Strait, given the end of the Soviet and Vietnamese threats and the reduction of concerns over India, and could simply overwhelm Taiwan's defences with a mass attack.[42]

On the other side of the Strait, Chinese leaders have at times expressed growing concerns over recent arms acquisitions by Taiwan, particularly foreign purchases such as the US F-16s and the French Lafayette frigates. Moreover, as indicated in the introduction, Beijing is becoming increasingly alarmed over the adverse implications for eventual reunification of both Taiwan's improved international standing and the domestic successes of separatist political elements within Taiwan, centred on the DPP.[43]

Although intended to support Taiwan's military acquisitions by creating an external political and economic environment that will strengthen deterrence, President Lee's foreign policy is viewed by many Chinese leaders as a thinly veiled attempt to pave the way for eventual formal independence.[44] Thus, Lee Teng-hui's strong reaffirmations of a 'one China' policy (reflected most notably in his establishment of a National Unification Council and its subsequent enunciation of a three-stage process of reunification) and his efforts to open an official government-to-government dialogue with Beijing are seen by the Chinese leadership as mere diversionary tactics. This belief became especially prevalent following Lee's June 1995 visit to the US. These observations and assessments together suggest several reasons, from Beijing's perspective, for increasing its military capabilities against Taiwan and for heightening the credibility of its threat to use direct force against the island. Thus, dynamics on both sides of the Strait at least suggest that a strong, interactive sense of threat or concern is driving the pattern of military acquisitions outlined above, leading to a full-blown arms race.

However, such a conclusion is probably premature, if not entirely inaccurate. As in the 1980s, one still cannot convincingly show that China's accelerated level of military acquisitions is aimed *primarily* at Taiwan, despite the applicability of some new weapons systems, technical capabilities and military exercises to that theatre of operations, and the growing concerns expressed by Chinese leaders. The underlying, broad-based doctrine for the development of China's overall naval and air capabilities remains essentially the same in the 1990s as it was in the 1980s, i.e. to bolster China's global prestige and regional influence, to contend with domestic and border unrest, to hedge against uncertainties regarding the future military and diplomatic posture of Japan, the United States and perhaps India, and to reinforce specific territorial claims, most significantly in the South China Sea.

The latter contingency in particular almost certainly remains the primary operational rationale behind the PLAN's efforts at improving its power projection capabilities. As suggested above, operations on or around the Spratly Islands require multi-faceted high seas naval capabilities of the sort most desired by the PLAN.[45] In particular, the attainment of force multipliers such as a mid-air refuelling technology, improvements in afloat support capabilities, and past exercises in or near the South China Sea that include the disruption of naval blockades all seem more applicable to operations in the Spratlys than against Taiwan.[46]

More broadly, one could argue that many of the above air and naval acquisitions are primarily intended by Beijing to counter similar military acquisitions recently undertaken or planned by several Southeast Asian states, some with claims to the Spratlys. In recent years, Singapore, Malaysia, Indonesia, and Thailand in particular have begun to develop or acquire modern weapons systems (e.g. strike aircraft, missiles, submarines, surface combatants, etc.) and force multipliers (e.g. air-to-air refuelling, enhanced surveillance systems, and airborne early warning aircraft) that present the potential of radically changing the maritime security environment in the South China Sea. Many of these countries are also developing limited rapid reaction forces.[47] All of these activities can be seen as providing significant motivation for many aspects of Beijing's military modernization effort.

Finally, if Beijing were focusing its military attentions primarily on Taiwan, one would expect to see signs of major efforts to prepare for the type of military operation that could directly threaten the island's security, i.e. an invasion. Yet even Taiwanese analysts acknowledge that the Chinese are not currently planning an amphibious invasion of Taiwan.[48] China does not regularly practise such large-scale operations, and even if it did, it is unlikely that effective naval gunfire would be available to support them. The PLAN also lacks a sufficient number of blue water transport and amphibious assault ships to launch a credible invasion. Specialists estimate that it would take the PLA several years of training and greatly expanded exercises to develop such a capability.

Thus, despite the new acquisitions of the 1990s, one should not assume that a full-scale arms race is now underway between Taiwan and China[49] or that the PLA is preparing for a massive attack against Taiwan. However, even if the above factors cast doubt on the use of the term 'arms race' to describe the state of Taiwan–China military relations, one must ask whether China is developing capabilities, for whatever purpose, that could produce a decisive advantage for Beijing and thus drastically increase its incentives to apply direct military force against Taiwan. A closer look at Chinese capabilities suggests that such a possibility is unlikely, at least over the medium term, although other less direct types of provocative military behaviour are possible.

A BALANCE OF FORCES OVER THE MEDIUM TERM

Despite their increasing capabilities, Beijing's forces still display severe weaknesses in critical areas necessary to any direct use of force against Taiwan. First and foremost, by most accounts, the PLA continues to suffer from very poor command and control capabilities for its air, naval, and land forces (including RRUs), despite recent improvements. As many analysts have pointed out, such a fundamental problem can greatly diminish the potential advantage presented by Beijing's superiority in numbers.[50]

Second, despite the above-noted improvements, by the end of the 1990s it is probable that few of the PLAN's principal surface combatants will be as modern or powerful as Taiwan's greatly improved fleet of nearly 40 US and French-designed frigates and destroyers.[51] Of particular note, Taiwan's surface combatants have more integrated fire control and countermeasures systems and both ASW and anti-aircraft capabilities than do their PLAN counterparts.[52] Also, some analysts believe that in actual blockade operations against Taiwan, Beijing may only be able to deploy a relatively small portion of its major warships on the high seas around Taiwan at one time for a sustained period, due to the limitations of nearby port facilities and the PLAN's lack of a redundant underway replenishment capability.[53] Moreover, without significant land or carrier-based air cover, China's destroyers and frigates would be highly vulnerable to missile or fighter attacks. All of these vessels lack a long-range SAM system and effective defence against anti-ship missiles.

Third, as suggested above, virtually all of China's submarines are of obsolete design and thus relatively easy to detect by modern sonar equipment. Also, most of China's five nuclear-powered attack subs, launched in 1972 and 1977, are probably no longer operational, and some may even have been scrapped.[54] Such a deficiency in submarine capabilities would also severely limit China's ability to maintain a strong blockade around Taiwan. Moreover, it will take many years for China to obtain and effectively utilize a significant number of improved or new subs, such as the Ming-class, Song-class, and the Russian Kilo-class subs noted above.

Fourth, China's large number of fast attack craft and minesweepers are defensive in nature, with limited operational range. Given their small size, many would probably encounter difficulty operating effectively on a rough sea around Taiwan.[55]

Fifth, as suggested above, the PLAAF still suffers from a host of difficulties, including a very weak air-to-ground attack capability, a generally poor missile inventory, insufficient combat training and logistical support, poor C^3I capabilities, a largely defensive fighter force structure, and overall obsolescence in airframe design and key technologies. The fighter bomber fleet has a combat radius of only about 280 nautical miles, largely insufficient for adequate air cover over the east coast of Taiwan, and the PLAAF is probably still several years from developing a fully operational and extensive aerial refuelling capability. Also, PLAN pilots are probably the only airmen trained to navigate over water at extended range.[56]

The acquisition by China of only another 24 or 48 Su-27s or comparable aircraft during the remainder of the decade will certainly not resolve the above problems.[57] Moreover, it will take many years for China to attain significant production levels for the Su-27 and, even at peak production levels, it is estimated that China will probably produce only about 50 aircraft per year through the co-production arrangement with Russia.[58] Equally important, it will take pilots, ground crews, and logistics personnel many years to master the required technologies and operational features of the Su-27. In the meantime, Taiwan's F-5Es are generally equal to China's J-7s, and better than their J-6s, and the overall quality of Taiwanese pilots is considered to be far superior to those of the mainland. Moreover, acquisition of F-16s and Mirages will give Taiwan a significant qualitative edge over most of the existing or planned PLAAF fighter force. Finally, these sales have opened the door for future additional Taiwanese aircraft acquisitions from foreign sources.

One should also mention that studies conducted by RAND researchers of PLAAF fighter production and retirement/attrition rates suggest that if the Chinese intend to keep their fighter force at current levels, over 1,500 additional aircraft would probably be required between 1995 and 2005. Assuming that J-8 production could be more than doubled, to 40 units per year, J-7 production would still need to exceed 140 per year, a major increase over current levels. Moreover, even if successful, the resulting fighter force would be composed of almost 60 per cent J-7 and J-8 aircraft, capable of intercepting aircraft over China, but still inadequate for power projection and ground attack especially without an extensive aerial refuelling capability.[59]

Sixth, China's huge air force is also greatly constrained by its limited support facilities. Taiwanese analysts have estimated that Chinese airfields adjacent to Taiwan can only hold 1,200 of China's nearly 6,000 combat aircraft. Of this number, they calculate that no more than 200 aircraft could be launched against Taiwan at one time, with a maximum of three sorties per day.[60] Moreover, air power specialists doubt that China could maintain even this tempo for more than a few weeks, given the relatively poor state of Chinese logistics and maintenance facilities. Initial PLAAF sorties, as well as possible missile attacks from the mainland, would probably be aimed at eliminating a large part of the Taiwanese air force. However, for this reason, Taiwan has constructed a huge tunnel complex of aircraft shelters and operational facilities for up to 300 fighters inside a mountain base on the east coast of Taiwan.[61]

The previous two factors together suggest that it would be extremely difficult for the PLA to establish and maintain air superiority over the waters surrounding Taiwan at any time during at least the remainder of the decade. Yet such an accomplishment is deemed by most specialists as necessary to carry out an effective blockade of the island or to support an invasion.[62]

Seventh, even Chinese improvements in stand-off attack capabilities will likely prove insufficient to tilt the balance decidedly in Beijing's favour. There are reports that China's efforts to improve the mobility and accuracy of its surface-to-surface and air-to-surface ballistic missiles are aimed in part at the development of a capability to surgically remove Taiwan's major military targets without producing extensive collateral civilian damage. Such an attack would ostensibly constitute a potent demonstration of Chinese military capabilities that would minimize the chances of rapid escalation, demoralize the Taiwanese public and convince Taiwan's leadership of the futility of armed resistance. However, it is difficult to believe that an attempted missile attack would produce such effects. First, it would probably take many months for China to develop and deploy a sufficient number of such precise missiles. This would almost certainly be detected by Taiwan's improved early warning and surveillance capabilities, thus giving the island time to implement effective countermeasures, probably with US assistance. Second, Chinese missiles almost certainly could not penetrate Taiwan's east coast air defence complex and thus would probably not prevent a strong counterattack by Taiwan. Such a counterattack would virtually ensure escalation to a full-scale, general conflict, which the mainland might not win. Hence, it seems unlikely that Beijing would be willing to gamble everything on the highly doubtful possibility that missiles alone will force Taiwan to capitulate.[63]

Taken together, the above mix of military capabilities and shortcomings suggests that the balance of forces between Taiwan and China is relatively stable and will probably remain so for many years, despite the improvements already implemented or planned by Beijing. Moreover, even if China were to obtain a slight edge in a particular capability, this would almost certainly prove insufficient to change the basic military calculus. Chinese leaders undoubtedly recognize that a military failure against Taiwan will result in the island's formal declaration of independence.[64] Hence, barring a strong provocation from Taiwan (see below), they will probably not want to launch an attack unless they are convinced that the PLA enjoys a clear military superiority, not simply a marginal edge.

Beyond such purely military factors, however, a range of political and economic disincentives also exists against the employment of direct force by Beijing, especially over the short and medium term (i.e. five to ten years). First and foremost, the Chinese leadership is fully aware that any attack on Taiwan would greatly threaten, if not entirely overturn, the reform-based domestic and foreign policy strategy that has been the key to China's diplomatic and economic successes for nearly twenty years, and in general destabilize the entire Asia–Pacific. Specifically, such a Chinese attack would:

1 Destroy the Sino–American détente that has underlain the stability and security of East Asia since the 1970s, increase the US military presence in Asia, and perhaps even precipitate armed conflict between the US and China.
2 Cause virtually every Asian country to doubt China's avowed commitment to the peaceful resolution of its territorial and other disputes with its neighbours, thus probably leading to a major acceleration in regional military acquisitions and other destabilizing types of behaviour.

3 Severely destabilize Hong Kong, thus producing adverse economic consequences for China and a crisis in Sino–British relations.
4 On a broader level, disrupt trade and investment flows throughout the region.
5 Undermine growth and reform in China's coastal provinces, and perhaps in China as a whole.
6 Probably generate political and social unrest within China.[65]

Because of the above military and non-military considerations, most security experts (including many on Taiwan) believe that there is little chance of a direct military conflict between the two sides for the remainder of the decade.[66] In sum, China almost certainly does not have the strength to invade Taiwan over that period.[67] It also is probably not capable of maintaining an effective blockade of the island, and almost certainly could not do so if the USA were to provide military assistance to Taipei, as seems likely in the event of an unprovoked attack by Beijing.[68] Moreover, many analysts believe that, to be truly credible, any blockade of Taiwan, as well as a direct missile attack, would need to be backed by a credible threat of invasion.[69] Hence, China's poor capabilities in the latter area add greatly to the existing disincentives for mounting the former operations.[70] As a result of these calculations, Taiwan's military planning for the latter half of the 1990s has shifted somewhat, away from efforts to counter an invasion to a stress on surveillance and early warning functions and the maintenance of air and naval capabilities necessary to prevent a blockade of the island.[71]

LONG-TERM PROSPECTS: THE IMPORTANCE OF NON-MILITARY FACTORS

What are the chances that existing military trends will lead to direct conflict or radically increased military tensions over the longer term, however? Such an assessment presents enormous difficulties, given the highly dynamic changes occurring on both sides of the Strait. Here we can only present some general observations. Assuming current GNP growth rates and defence spending levels, at least one estimate suggests that Chinese military expenditures could exceed US$125 billion by the year 2007 (four to five times current estimated levels), while Taiwan might be spending only US$23 billion on defence.[72] As Table 10.2 suggests, such a level of Chinese spending would probably transform the military balance across the Strait, giving Beijing a decisive edge.

Table 10.2 PPP-based Chinese defence spending projection (in US$ millions)

	1992	*1997*	*2002*	*2007*
China	38,133	53,003	81,482	125,261
Taiwan	8,628	12,614	17,281	23,675

The chances of such a pessimistic outcome would increase further if growing Chinese concerns over domestic political developments on Taiwan actually led by that time to the commitment of an increasing proportion of Chinese defence funds to the improvement of specific military capabilities intended for use against the island. Also, by the year 2007, the Chinese leadership may have fully weathered the turmoil of the succession process and, with continued economic growth and domestic stability, attained a level of self-confidence that might embolden them to defy the international community and apply overwhelming direct military force against Taiwan. Taken together, these factors suggest that the Chinese threat to the island could well increase greatly over the long term.

However, such pessimistic, 'straight-line' extrapolations of existing trends should be viewed with a great deal of scepticism. First, projections of future Chinese defence spending are highly unreliable and thus provide a weak basis for assessing the long-term development of the Taiwan–China military balance. The above projection of Chinese defence spending employs a purchasing power parity (PPP) methodology, which results in an extremely high projection. Other higher and lower PPP-based estimates are also possible, as well as several non-PPP-based calculations. For example, a radically different projection would result if one were to use a more modest, non-PPP methodology derived from US government sources, as shown in Table 10.3.[73]

Table 10.3 Non-PPP-based Chinese defence spending projection (in US$ millions)

	1992	*1997*	*2002*	*2007*
China	15,800	23,910	33,530	47,030
Taiwan	8,628	12,614	17,281	23,675

This projection suggests that, although significant, Chinese military capabilities will not overwhelm those of Taiwan. In fact, the proportion of defence spending between Beijing and Taipei remains very similar in 2007 to what it was in 1992, at around 2:1.

Second, one should not assume that the Chinese leadership will be able to continue existing levels of defence spending over such a long period. The Chinese government faces a major, and growing, fiscal crisis, as public revenues decline and expenditures mount.[74] Beijing will need to carry out a radical restructuring of its tax system in order to remedy this problem and ensure high levels of military funding over a sustained period. Although it is currently moving in such a direction, the obstacles remain enormous, and the effort could easily fail, thereby probably ensuring a steady decline in central government capabilities.[75] In contrast, Taiwan's government enjoys a well functioning tax system and will probably be able to support considerably increased levels of military spending, and thus may be able to 'keep up' with the mainland.

Third, with economic progress and political stability, incentives will probably grow, not diminish, for China to resolve the Taiwan issue peacefully. Continued Chinese growth will almost certainly bring even closer economic, cultural, and diplomatic ties with the Asia–Pacific, thus strengthening élite interests within and outside China favouring greater cooperation. More specifically, bilateral economic and other ties between Taiwan and China will probably continue to expand at a rapid rate, increasing the costs to both sides of a future conflict. Moreover, Taiwan's growing regional and global contacts will undoubtedly further constrain Beijing's ability to pressure other countries to acquiesce in a more coercive approach to the island.[76]

The greatest threat to the peaceful resolution of the Taiwan issue is not posed by existing or likely future military trends or the natural emergence of China as a more confident and militarily capable power. Instead, conflict would most likely result, over both the long and short term, from major leadership miscalculations, largely arising from domestic factors at work on either side.

On the Taiwan side, the most serious miscalculation would consist of a declaration of formal independence or some lesser action(s) viewed by the Chinese as clear preparation for independence. Most analysts believe that Beijing would almost certainly use military force against Taiwan under such circumstances.[77] Moreover, such a Chinese resort to force would probably occur regardless of the state of the military balance at the time or the adverse consequences such action would pose for Chinese reform policies and Beijing's relations with other powers.[78] The most likely prelude to such a Taiwanese miscalculation would probably be the emergence to power of the DPP. However, the recent crisis precipitated by Lee Teng-hui's June 1995 visit to the USA, which prompted PLA missile 'tests' in the vicinity of Taiwan and a series of threatening PLA military manoeuvres, suggests that serious miscalculations can also be made by President Lee Teng-hui, resulting from his search for greater international recognition for Taiwan (more on this latter point below).

On the Chinese side, a miscalculation would most likely consist of Beijing's sudden or gradual shift to a more threatening diplomatic stance and an eventual resort to force, perhaps beginning with provocative military displays in the Strait or other intensified attempts at armed intimidation. Beijing would probably seek to explain such actions by pointing to the leadership's loss of patience over Taiwan's continued efforts to avoid reunification negotiations or to perceived indications of a more deliberate move by Taipei toward independence. However, if Taiwan had done nothing new to greatly escalate tensions between the two sides, and if most of the above military and non-military disincentives against the use of force were still applicable, it is more likely that such dangerous Chinese behaviour would occur as a result of domestic factors. These might involve attempts by the élite to play the 'nationalism' card for political ends. For example, a confrontation with Taiwan could serve to unify a fragmented and conflictual successor leadership, distract popular attention from internal woes such as an economic crisis, or strengthen military support for (or control over) a weak, insecure post-Deng Xiaoping regime. Such a confrontation could also be used as a gambit by individual groups or factions within the broader civilian or military leadership to further their narrow political or institutional interests. For example, the PLA Navy might seek a more assertive stance towards Taiwan in order to justify stronger central support and bigger budgets for the development of a modern, technically–proficient, combat-ready, blue water capability, as it has done with policy toward the Spratlys. The ability and opportunity of the PLAN to take such an action could arguably increase greatly in a post-Deng setting.[79]

Prior to the events of June–August 1995, the chances were comparatively greater that Beijing, not Taipei, would miscalculate or take major risks that could prompt a serious military conflict. First, the Chinese threat to use force in the event of *unambiguous* moves by Taiwan toward independence was clear and explicit. In other words, Taiwan knew what actions would trigger a violent Chinese reaction.[80] Second, the emergent electoral process on Taiwan placed major checks on the ability of any Taiwanese government to make bold policy shifts, such as a declaration of independence. Third, DPP politicians were increasingly divided over whether, when and how Taiwan should declare independence from the mainland, and they will likely remain divided over these issues even if their party were to attain power. Fourth, the

DPP needed to be in full control of both the executive and legislative branches of government to make an independence declaration effective and politically compelling. Moreover, the emergence of genuine levers for the expression of opposition sentiment would have allowed the KMT and other parties such as the New Party to resist or block any DPP moves toward formal independence.[81]

Most of these factors still hold true today. However, the events of June–August 1995 suggest that President Lee Teng-hui is willing to test the extreme limits of Beijing's tolerance and thereby risk a direct military conflict. Beijing's unprecedented military response indeed suggests that Lee seriously miscalculated. The threshold for the future use of force against Taiwan may have been significantly lowered as a result, thus altering Beijing's overall calculus. As indicated above, the Beijing leadership now believes Lee Teng-hui is committed to the independence of Taiwan and must be deterred by a more potent threat of military force. Even more alarming, a growing number of senior officers within the PLA reportedly believe Japan and perhaps the United States are secretly encouraging Lee in his search for independence, and that Beijing will *inevitably* be forced to attack and take the island before it is lost altogether.[82]

Thus, the restraints on a Chinese miscalculation are now arguably fewer and weaker than prior to June 1995. Moreover, China's uncertain domestic situation could provide increased opportunities and incentives for the type of provocative actions toward Taiwan outlined above. Chinese leaders in the past have taken major external risks and used or provoked confrontations with other states for domestic political ends. One should not assume that such dangerous behaviour will disappear, even if the communist regime collapses in China. Finally, Chinese leaders might believe that the US threat to intervene in a China–Taiwan conflict has lost much of its credibility in recent years. If true, this could further increase Beijing's inclination to use a confrontation with Taiwan for domestic political purposes.[83]

Despite the arguably increased likelihood of miscalculation by either Beijing or Taipei, however, it must be reaffirmed that the above-mentioned military, political and other disincentives working against conflict between the two sides still remain valid. Indeed, the costs to both sides of a military clash will probably increase, not decrease, over time. The challenge for both sides, and for the United States, is to develop policies and practices that increase mutual incentives to avoid conflict while lowering the chances for miscalculation. This could prove to be the most essential, and difficult, task for ensuring continued peace and stability in East Asia.

APPENDIX

Table 10.4 Military balance, 1995: China and Taiwan

		China		*Taiwan*
Size of armed forces		3,030,000		442,000
Size of reserves		1,200,000		1,657,500
Strategic nuclear force	ICBMs	14		none
	IRBMs	60		none
	SLBMs	12		none
Size of army		2,300,000		240,000
Number of main battle tanks		7,500–8,000		570
Combat aircraft[1]		5,000+		≈1,000
Bombers		630		none
	H-6	150[2]		
	H-5	480[3]		
Fighters		4,000+[4]		800+[5]
			F-5	275
	J-5	400	F-5B	8
	J-6	3,000	F-5E	215
	J-7	500	F-5F	53
	J-8	100	F-104 D/DJ	8
	Q-5	500	F-104/G	81
	Su-27	26	F-104/J	20
			TF-104G	32
			IDF(Test)	10
Helicopters		≈380		≈260
Naval	(ASW): SA-321	15	(ASW): Hughes 500 MD MSW	12
	Z-5	40	2H-2F	12
	Z-9	10	S-70B	10
Transport (Air Force)		600		8 SQN
Principal surface combatants		50		38
Destroyers		18		22
DDG[6]		18		8
DD[7]		none		14
Frigates		32		16
FFG[8]		33		4
FF[9]		5		12
Submarines (diesel)		47[10]		4
SSBN[11]		1		none
SSN[12]		5		none

Table 10.4 Military balance, 1995: China and Taiwan (cont.)

	China	*Taiwan*
Patrol and coastal combatants	870	98
Mine warfare ships	≈ 126	15
Mine sweepers (coastal)	93	9
Minesweepers (ocean)	41	none
MCM (Mine countermeasures ships)	≈ 125	15
Amphibious ships	51	26
Amphibious landing craft	400+	400

Notes:
[1] Including Naval aircraft
[2] Includes 30 H-6 PLAN
[3] Includes 130 H-5 PLAN
[4] Includes 700 PLAN (J-5, J-6, J-7, J-8, Q-5)
[5] Includes 32 Taiwan Navy
[6] Destroyer with area SAM
[7] Destroyer
[8] Frigate with area SAM
[9] Frigate
[10] Probably about additional 50 Romeo-class non-operational
[11] Nuclear-fuelled ballistic missile
[12] Nuclear-fuelled

Sources: *The Military Balance, 1995–1996*, London, October 1993; Sharpe, Richard (ed.) *Jane's Fighting Ships 1993–1994*, Coulsdon, 1993

NOTES

1 The other major threat is of course posed by the possible acquisition of nuclear weapons by North Korea.
2 See Colin S. Gray, 'The Arms Race Phenomenon', *World Politics*, 24, no. 1, October 1971, pp. 39–79.
3 The following analysis of arms expenditures and trade levels is based on figures provided by the Stockholm International Peace Research Institute and the US Arms Control and Disarmament Agency. It must be stressed that these data (and similar statistics presented in other sources) are often subject to significant margins of error. Hence, they are mainly used to indicate *general trends*, not to provide precise estimates of yearly levels.
4 While the Chinese government claimed a figure of less than 2 per cent of GDP during the 1980s, US intelligence sources believed the amount was closer to 3.5 per cent. See the discussion in *SIPRI Yearbook 1992*, Oxford, 1992, p. 248.
5 Taiwan was the second largest arms importer in Northeast Asia during the 1980s, second only to Japan, which purchased nearly US$10 billion in

weaponry. See *World Military Expenditures and Arms Transfers 1990*, US Arms Control and Disarmament Agency, Washington, 1991.

6 Gary Klintworth, 'Taiwan: Emerging Actor in Asia–Pacific', *Asia–Pacific Defense Reporter 1994 Annual Reference Edition*, December 1993/January 1994, p. 54.

7 The August 1982 Sino–US Joint Communiqué provided the impetus for the steady reduction of US arms sales to Taiwan during the remainder of the 1980s. Sales decreased by about US$20 million per year, beginning in 1983. See Gary Klintworth, 'Taiwan: Leading From Strength', *Asia–Pacific Defense Reporter 1993 Annual Reference Edition*, December 1992/January 1993, p. 51.

8 John W. Garver, 'China's Push Through the South China Sea: The Interaction of Bureaucratic and National Interests', *China Quarterly*, no. 132, December 1992, pp. 999–1028. As Garver argues, in the 1980s, the PLAN under General Liu Huaqing used Chinese territorial claims over the South China Sea in order to justify the need for a sophisticated, high-seas power projection capability and hence larger defence budgets. This effort arguably continues today.

9 For details, see Parris Chang, 'Beijing's Relations with Taiwan', in Parris H. Chang and Martin L. Lasater (eds) *If China Crosses the Taiwan Strait: The International Response*, Maryland, 1993, pp. 3–13. For detailed information on the expansion of economic relations between the two sides, see Robert F. Ash and Y.Y. Kueh, 'Economic Integration within Greater China: Trade and Investment Flows Between China, Hong Kong, and Taiwan, *China Quarterly*, no. 136, December 1993, pp. 711–45; and Chong Pin-Lin, 'Beijing and Taipei: Dialectics in Post-Tiananmen Interactions', ibid., pp. 779–85.

10 Andrew Mack and Desmond Ball, 'The Military Build-up in Asia–Pacific', *The Pacific Review*, vol. 5, no. 3, 1992, p. 197. Chong-Pin Lin, December 1993, states (p. 793) that Taiwan's defence budget increased 27 per cent in 1990 and 7 per cent in 1991. The budget was over US$10 billion in 1992.

11 Andrew Mack and Desmond Ball, 1992, p. 200.

12 The official Chinese defence budget increased over 15 per cent in 1989 and 1990, nearly 14 per cent in 1991, over 14 per cent in 1992 and 1993, and by 23 per cent in 1994. For reference to planned Chinese spending levels, see *South China Morning Post*, 7 March 1991, p. 1.

13 Most of the following information on Taiwan's recent and planned military acquisitions was obtained from: *The Military Balance, 1993–1994*, and *1994–1995*, London, 1993 and 1994, including the enclosed fold-out wall chart entitled 'Asia: The Rise in Defence Capability, 1983–1993'; Gary Klintworth, December 1992/January 1993, December 1993/January 1994, and 'Developments in Taiwan's Maritime Security', *Issues and Studies*, vol. 30, no. 1, January 1994, pp. 65–82; A.W. Grazebrook, 'The Year at Sea: Regional Naval Growth Continues', in *Asia–Pacific Defence Reporter 1994 Annual Reference Edition*, vol. 20, no. 6/7, December 1993/January 1994, pp. 6–7, and A.W. Grazebrook, 'The Year at Sea: More Regional Naval Growth', in ibid., 1995, pp. 12–13; Joseph R. Morgan, *Porpoises Among the Whales: Small Navies in Asia and the Pacific*, East–West Center Special Report, no. 2, March 1994; Victor Fic, 'Taiwan's Air Force: Redefining ROCAF and National Defence', *Asian Defence Journal*, August 1993, pp.

43, 44, 46, 48; Chong Pin-Lin, December 1993, pp. 789–99; Julian Baum, 'Steel Walls', *Far Eastern Economic Review*, 9 July 1992, pp. 9–11, and Dennis Van Vranken Hickey, *United States–Taiwan Security Ties: From Cold War to Beyond Containment*, Westport, Conn., Praeger Press, 1994, pp. 41–93.

14 Julian Baum, 9 July 1992, p. 11.

15 This may present a major problem, however, since Beijing has thus far managed to deter the intended Dutch and German suppliers from completing the sale.

16 Gary Klintworth, December 1993/January 1994, p. 54.

17 The A and B versions are less capable than the C and D versions. However, Taiwan will reportedly upgrade packages from General Dynamics by the time of first delivery in 1996. See Julian Baum, 'A Foot in the Door', *Far Eastern Economic Review*, 17 September 1992, p. 13.

18 Gary Klintworth, December 1993/January 1994, p. 54.

19 'Army is "Beefing Up" Development Capacity', Taipei, China News Agency, 25 January 1995, in FBIS–CHI, 26 January 1995, p. 97.

20 Victor Fic, pp. 46, 48.

21 Most of the following information on China's recent or planned military acquisitions was obtained from: *The Military Balance 1993–1994* and *1994–1995*; Larry Wortzel, 'Engaging the New World Order: Where is the PLA?' paper prepared for the Fourth Annual Staunton Hill Conference on the PLA, 27–9 August 1993; Larry Wortzel, 'China Pursues Traditional Great-Power Status', *Orbis*, Spring 1994, pp. 157–175; Chong-Pin Lin, December 1993, pp. 789–99; A. W. Grazebrook, 1993 and 1995; Joseph R. Morgan, 1994; Yann-Huei (Billy) Song, 'China and the Military Use of the Ocean', *Ocean Development and International Law*, vol. 20, 1989, pp. 213–35; Desmond Ball, 'Arms and Affluence: Military Acquisitions in the Asia–Pacific Region', *International Security*, vol. 18, no. 3, Winter 1993/1994, pp. 78–111.

22 China's most modern and largest warship is the 4,200 tonne Luhu destroyer. One of an expected 4 Luhu's had been built by early 1995. The improved Luda III destroyer is intended to replace 16 earlier and largely obsolete versions. As many as 6 Jiangwei frigates could be in service by the end of 1995, joining or partly replacing China's fleet of approximately 30 ageing Jianghu-class frigates.

23 The PLAN has added over 100 mine warfare ships to its inventory since the mid–1980s.

24 Overall, a sixth improved Ming-class submarine and one modified Romeo-class with an Exocet-type surface-to-air missile have been commissioned thus far.

25 There are also reports of negotiations with Russia to undertake the licensed production of Kilo subs, which are well-suited for coastal waters and for operations such as naval blockades. See Kathy Chen, *Asian Wall Street Journal*, 9 February 1995.

26 *The Military Balance 1993–1994*, p. 148; Colonel John Caldwell, *China's Conventional Military Capabilities, 1995–2004: An Assessment*, The Center for Strategic and International Studies, Washington DC, 1994; and Karl W. Eikenberry, 'Does China Threaten Asia–Pacific Regional Stability?' *Parameters*, vol. 25, no. 1, Spring 1995, pp. 83–103.

27 Larry Wortzel, 1994, p. 164.
28 Yann-Huei (Billy) Song, 1989, p. 226. Also see John Caldwell, 1994, p. 8.
29 Larry Wortzel, 1993, p. 21. Wortzel states that the H-6 is only produced at a rate of four per year.
30 Ibid., p. 17. According to Wortzel, systems integration and engine design and manufacturing are the key obstacles.
31 *The Military Balance 1992–1993*, p. 148.
32 Wortzel states, ibid., p. 17, that since 1991 over 1,000 Russian defence scientists have reportedly conducted defence industrial exchanges with China.
33 China has also purchased 10 IL–76 medium range transport aircraft from Russia, mainly to improve the mobility and lift capability of China's rapid reaction units (see below for more on these units).
34 Larry Wortzel, 1994, p. 169.
35 Desmond Ball, 1994, p. 103.
36 Larry Wortzel, 1993, p. 23. Given the size of its merchant marine, Wortzel estimates that China could probably move the equivalent of a full Group Army (about 40,000 men) to conduct follow-up landing operations after the seizure of a beachhead.
37 Ibid., pp. 24–5.
38 Paul H. B. Godwin, 'The Use of Military Force Against Taiwan: Potential PRC Scenarios', in Parris H. Chang and Martin L. Lasater (eds), 1993, p. 20.
39 Ibid., 'Conclusion', pp. 170–171.
40 See the reports on the Defence White Paper in the Taipei CNA report, 9 March, 1994, in FBIS 94–046, 9 March, 1994, p. 69, and the Hong Kong AFP report, 15 March 1994, in FBIS 94–050, 15 March 1994, p. 64.
41 For example, see the remarks of Admiral Liu Ho-chien, Taiwan's Chief of the General Staff, in ibid., 15 March 1994 and in FBIS-CHI-94–049, 14 March 1994, p. 80. This window of opportunity for Beijing will supposedly last until at least 1996, when Taiwan takes delivery of significant numbers of its advanced foreign fighters and warships.
42 Parris H. Chang and Martin L. Lasater (eds), 1993, 'Conclusion', p. 171, and Paul Godwin, 1993. Godwin states (p. 27) that 'without support from the United States, Taiwan's forces would be unable to conduct an effective defense against invasion'.
43 President Lee Teng-hui's multi-faceted strategy of 'flexible diplomacy' has yielded many benefits for Taiwan since its inauguration in the late 1980s. It has expanded Taipei's ties with the world and thus apparently increased regionwide and global interest in a peaceful resolution to the Taiwan issue. Specifically, Taiwan is again participating in major international organizations such as the Asian Development Bank, has improved links with many countries, including the former communist states of Eastern Europe as well as Russia and several of the former Russian republics, has established formal diplomatic ties with several small countries, and in general enjoys expanding economic relations with an increasing number of countries, including Vietnam and many members of ASEAN. Within Taiwan, President Lee's support for greater levels of democratization has produced a strong and irreversible trend toward the Taiwanization of the political system, marked, among other things, by the increasingly public

discussion of options for the establishment of formal independence and the rise in power of the separatist-minded DPP. For further details, see Parris Chang, 1993; Michael Yahuda, 'The Foreign Relations of Greater China', *China Quarterly*, no. 136, December 1993, pp. 687–710; and Yu-Shan Wu, 'Taiwan in 1993: Attempting a Diplomatic Breakthrough', *Asian Survey*, vol. 34, no. 1, January 1994, pp. 46–54.

44 This point is made by Martin L. Lasater, 'Principles of Deterrence in the Taiwan Strait', in Parris H. Chang and Martin L. Lasater (eds), 1993, p. 158.

45 Therefore, as John Garver asserts (p. 1022), 'a close, symbiotic relationship (has) developed between the tasks the PLAN has to accomplish in the South China Sea and the progressive modernization of its long-distance, high seas capabilities'. Also see Joseph Morgan, p. 34, who states that PLA General Zhao Nanqi has explicitly remarked that China's naval and air modernization efforts are intended to bolster claims to the Spratlys and extend Beijing's military presence into the Indian Ocean.

46 This is of course not to deny that many of Beijing's naval and air exercises of summer and autumn 1995 have been specifically aimed at Taiwan, and designed to increase China's capability to wage coercive diplomacy against the island. More on this point below.

47 Desmond Ball, 1994, pp. 102–3.

48 The smaller amphibious operations noted above are far more applicable for the South China Sea.

49 Desmond Ball, 1994, p. 94, arrives at the same conclusion for the Asia–Pacific region as a whole.

50 For example, Godwin, p. 19, and Eikenberry, 1995, p. 87.

51 Gary Klintworth, 'China: Myths and Realities', *Asia–Pacific Defense Reporter*, April-May 1994, p. 14.

52 Larry Wortzel, 1993, p. 14.

53 Ibid., p. 15, A Hong Kong publication on the PLAN entitled *Chinese Communist Naval Forces*, Hong Kong, June 1993, argues that China could probably field only about 12 of their destroyers in ocean combat. See FBIS-CHI-002, 5 January 1994, p. 30.

54 Joseph R. Morgan, 1994, p. 33.

55 Ibid., p. 34.

56 Larry Wortzel, 1993, p. 17. Also see FBIS-CHI-002, 5 January 1994, pp. 30–31, and Paul Godwin, 1993.

57 According to Gary Klintworth, April–May 1994, p. 14, USAF intelligence specialists believe that the PLAAF's 24 Su-27s have little operational significance and that even 75 aircraft will have a relatively modest impact compared to Taiwan's aircraft acquisitions.

58 Tai Ming Cheung, *Far Eastern Economic Review*, July 8, 1993.

59 Interview, RAND researchers, and Kenneth Allen, et al., *China's Air Force Enters the 21st Century*, RAND, Project Air Force, MR-580-AF, 1995, pp. 163–165. This detailed study provides a comprehensive analysis of the past, present, and likely future features of the PLAAF. It also confirms the many obstacles faced by that service arm in its efforts to modernize.

60 Gary Klintworth, December 1992/January 1993, p. 50.

61 Gary Klintworth, ibid., p. 51.

62 Parris H. Chang and Martin L. Lasater (eds), 1993, 'Conclusion', p. 169, and Godwin, pp. 25, 29.

63 At least one knowledgeable observer of the China–Taiwan security relationship has told the author that a sudden, well-timed and focused missile attack by China would probably create enormous panic on the island, leading to a massive exodus and the acceptance of Beijing's diplomatic terms for reunification. Such an assessment largely assumes that the Taiwanese public believe that their island is indefensible against a mainland attack. If true, such (in my view incorrect) notions suggest that the Taipei Government may need to revise its public stance on this issue, more accurately to reflect the state of the military balance across the Strait.

64 Godwin, 1993, p. 16.

65 These and other factors are discussed by Gerald G. Segal, 'Chinese Strategy to the Year 2000', in *SCPS PLA Yearbook 1988–1989*, Taiwan; 1989, pp. 34, 35; Paul H. Kreisberg, 'Asian Responses to Chinese Pressures on Taiwan', in Parris H. Chang and Martin L. Lasater (eds), 1993, pp. 95–6, and ibid., 'Conclusion', p. 166.

66 Ibid. This analysis concludes (p. 173) that 'it is unlikely that the PRC will use force against the ROC'. Also see pp. 192–3. In fact, some Chinese leaders have suggested that they are willing to give Taiwan until at least the turn of the century to respond favourably to Beijing's 'one country two systems' proposal. See Martin Lasater, ibid., p. 159.

67 Ibid. The authors conclude (p. 175) that the Chinese could develop an amphibious invasion capability, if they dedicated enormous efforts to that end for the next 5–10 years.

68 For example, Martin Lasater, ibid. (p. 157) states 'it is highly probable that the United States would aid Taiwan if it were attacked by the PRC'. Also see Paul Kreisberg, ibid., p. 88.

69 This is another conclusion of ibid., p. 172.

70 Paul Godwin, ibid., presents in some detail the enormous difficulties that would face an attempted Chinese naval blockade and potential invasion.

71 See Gary Klintworth, January 1994, p. 75, and Julian Baum, 9 July 1992.

72 These calculations are based upon a steady but modest 7 per cent GNP growth rate for China, but a rather high average defence spending/GNP ratio of 5 per cent for 1987–92. For details, see Michael D. Swaine and Courtney Purrington, *Asia's Changing Security Environment: Sources of Adversity for US Policy*, RAND, DRR-655-3-1-OSD, July 1995.

73 In this case, Chinese calculations assume the same 7 per cent GNP growth rate and a slightly lower average defence spending/GNP ratio for 1987–92 of 4.1 per cent. A more extended discussion of the problems involved in estimating and projecting Chinese defence spending is contained in Swaine and Purrington, July 1995, pp. 136–138. This analysis is based, in large part, on an excellent paper produced by the IISS, entitled 'Discussion Paper on Chinese Military Expenditure', presented at the IISS/CAPS Conference on Chinese Economic Reform: The Impact on Security Policy, 8–10 July, 1994.

74 For example, see Christine Wong, 'Central–Local Relations in an Era of Fiscal Decline: The Paradox of Fiscal Decentralization in Post-Mao China', *China Quarterly*, no. 128, December 1991, pp. 691–715.

75 One alternative is to expand non-tax sources of revenue by increasing arms sales. This will probably not generate sufficient funds, however, and could precipitate both increased tensions with the West and an internal leadership dispute. Both consequences would then serve to distract China's attention from Taiwan.

76 Gerald Segal, 1989, p. 35.

77 Paul Kreisberg, 1993, p. 82 provides several examples of possible Taiwanese political actions that would probably drastically raise tensions or trigger a Chinese attack. Also see, ibid., 'Conclusion', pp.175–77 and Vernon V. Aspaturian, 'International Reactions and Responses to PRC Uses of Force Against Taiwan', ibid., pp. 140–2.

78 Indeed, the Chinese leadership might calculate that such a clear Taiwanese 'provocation' would weaken support among the US public and political élite for American intervention, thus further raising the likelihood of an armed reaction. Such a calculation may indeed be accurate. See Paul Kreisberg, 1993, p. 89.

79 For details on this last point, see Michael D. Swaine, *China: Domestic Change and Foreign Policy*, RAND, Santa Monica, MR–604–OSD, 1995.

73 It was (and remains) generally understood that Beijing would attack Taiwan if: a) Taiwan were invaded or controlled by a foreign power, b) Taipei developed nuclear weapons, c) Taiwan claimed to be an independent state, d) the KMT refused to negotiate reunification over an ill-defined 'long period of time'. For these points, see Parris H. Chang, and Martin L. Lasater (eds), 1993, 'Conclusion', pp. 175–6. However, one less certain 'trigger' is presented by China's position toward the Kuomintang (KMT). Beijing has at times indicated that the KMT's loss of power alone might constitute a sufficient provocation justifying the use of force against Taiwan. Yet while clearly originating with the Taiwan side, this can hardly be called a 'miscalculation'.

81 These and other points are made in a RAND analysis of the implications of domestic change in Taiwan for conflict with China. See Evan Feigenbaum, *Change in Taiwan and Potential Adversity in the Strait*, RAND, MR-558-1-OSD, 1995.

82 Interviews conducted by the author in Beijing, July 1995.

83 The potential exists for a fundamental miscalculation between Washington and Beijing over American willingness to intervene in a Taiwan–China clash. As Martin L. Lasater points out (1993, p. 157), the US commitment to Taiwan may actually be growing, not declining, due to Taiwan's democratization and increased importance as a trading partner, the lowering of China's strategic significance to the USA after the collapse of the Soviet Union, US anger over the Tiananmen Square incident of June 1989, Chinese human rights abuses and unfair trade practices, and Beijing's missile and nuclear technology sales to the Third World. At the same time, several strategists in Beijing have told the author that the Chinese Government believes the US would not risk the life of a single American soldier to protect Taiwan. To support this view, they point to the US's unwillingness to use (or persist in the use of) military force in places such as Bosnia and Somalia. Such a divergence of views between Washington and Beijing, combined with the reduced Chinese tolerance toward Taiwan's actions, arguably present the most likely source of a future Chinese miscalculation.

It is therefore necessary for the United States to effectively counter the notion that its commitment to a peaceful resolution of the Taiwan issue has in any way diminished, while also making clear to the Taiwan Government that it does not support President Lee's brinkmanship.

Part III

Conclusions

11 Tying China in (and down)

Gerald Segal

There is no more weighty uncertainty for East Asia than the future of China. If China should stagger through with leadership struggles and perhaps even a disintegrating state, then the region will worry about mass migration and spreading chaos. If China should power ahead with double-digit growth, East Asia will worry about the implications of Chinese power. Futurology has always been a mug's game, but senior fellows in prestigious think tanks like to believe that it is legitimate to try to 'anticipate' the future. Hence the following, central theme: no matter whether China is weak or strong (or anything in between), the outside world has an interest in tying China in to the international system. By so doing, the implications of either Chinese chaos or Chinese power will be managed by a worried world.

In the wave of (post-Tiananmen) relief-induced optimism about the future, the current fashion is to be optimistic. William Overholt regards China as the next economic superpower while Harry Harding is predictably more cautious.[1] There are only a few congenital Cassandras who worry about the risks of regionalism or see a China that has 'flunked the fundamentals' and consequently is out of control.[2] But most should be willing to agree on the following categories of change.

CHINA CHANGES SHAPE

Even the most myopic visitor to China is aware of the mainly peaceful revolution underway. Many changes can be described as part of a general decentralization of authority, only some of which is voluntary. The most commented upon decentralization is in economic policy. This is not the place to rehearse the story of the economic reforms, but what began as a relatively carefully controlled experiment with market forces, has now slipped out of the control of central planners. Less than a quarter of the Chinese economy is run under the state plan and less than half is owned by the central state. Richard Margolis suggests that the Chinese economy is up to two-thirds hidden from central statisticians. Beijing contributes a shrinking amount of total national investment and strives (and mostly fails) to control the tax system.[3]

As one might expect with market forces, the power that has flooded away from Beijing, has not gone to a single rival authority but rather to a range of other levels of authority. The result is a looser sort of China where authority is more diffuse. In many aspects of economic policy it is hard to talk of 'running China'. If the leaders in Beijing were true believers in market forces, they would have less trouble in accepting this diffusion of power. Though economic authority has ebbed, old political structures and habits of mind remain. Hence the recurrent attempts made in Beijing to regain control over the economy. To be sure, such attempts are also encouraged by the likes of the World Bank and some foreigners who would prefer to deal with one authority in China. But the task of regaining the past system is long since lost. Chinese leaders and foreigners alike are coming to terms with the existence of many Chinese economies. One World Bank study notes that inter-provincial trade is shrinking relative to provincial trade with the outside world, and that provinces are increasingly taking on the attributes of independent actors in the international economy.[4]

The uncontrolled growth has also meant that investment in infrastructure (freight, energy, financial services) is only half the level of growth in industrial output. The result is an increasing need to trade with the outside world, and not merely a tap of desire that can be turned off at will. The shambolic state of domestic financial services has led to uninvested savings of some US$250 billion and a severe financial shortage for a central state increasingly desperate to fund 'brain dead' state industries. Chinese exporters are much more reliant on foreign capital than their counterparts elsewhere in East Asia. Foreign direct investment (FDI) accounted for less than 5 per cent of Chinese output, but this foreign investment was the source of two-thirds of Chinese exports. The bulk of Chinese exports is only loosely linked to the growth-through-trade mechanisms that have been so successful elsewhere in East Asia.[5]

It would be remarkable if this fundamental shift in economic power did not also affect other dimensions of policy. The decentralization of power has led to political leaders at all levels taking up opportunities to make money. The blurring of private and public function leads to corruption and simply bad government. The absence of an impartial legal system means there are few limits on the unscrupulous and hot money flows out of the country in search of security. In short, social order is increasingly at risk. The changes are staggering in their proportion and carry with them much revolutionary potential. There are at least one hundred million Chinese on the move from countryside to urban centres. Evidence suggests that at least an equivalent number of people are on the verge of making the same move. Such trends are common in modernizing peasant societies, but the scale of the Chinese population means there are more millions flooding into cities.[6] The rapidity of change makes it all the harder for people to adjust their mindsets and the strain is reflected throughout society. As long as the economy prospers a large part of the growing expectations of the people on the move can be met. But should growth slow or halt, there is revolutionary tinder scattered across the land.

The rapid social change is evident in myriad ways – beggars on streets, Dickensian labour practices, rapidly increasing drug addiction, gun running, gangsterism, rampant corruption, secret societies – the list seems endlessly bleak. Major crime rose 18 per cent from 1992 to 1993, and even incendiary devices exploded in major cities in the spring of 1994.[7] Economic growth buys off much discontent and while it continues, the system can continue in its anarchic way. But what if growth stops? For the time being, these changes are merely startling for those who remember a China of strict control of population and a relatively rigid social order. Authorities in the centre and regions try to control these social forces, but they struggle in vain. Foreigners may wish to see less migration spilling beyond the frontiers, fewer problems with piracy, and even safer streets and less corruption in business dealings, and Chinese leaders will no doubt concur. But the reality is that no level of authority in China specifically controls these problems. Piracy or mass migration is not a function of a clever central government trying to make mischief for the outside world, but it serves the interests of many in the outside world and in China to pretend they can control these processes.[8]

Related to the economic and social change is the major damage to the environment. China has 20 per cent of the world's population but only 7 per cent of its arable land. The ratio of arable land to worker is less than that of Bangladesh. Not surprisingly, rapid growth in farm output has ended earlier in China's development than elsewhere in East Asia. The loss of arable land is on a staggering scale, for even the building boom leads to bricks being built out of fertile land that once fed the cities. China is on an irrevocable course to becoming one of the world's largest food importers. Demographics demonstrate that China has no alternative but to depend on the outside world for basic sustenance.

China is no different from any other rapidly developing country in its propensity to ravage its environment, but the scale and speed of growth makes the ravaging all the more savage. More than half the Chinese population is said to drink water below World Health Organization (WHO) standards and major Chinese cities have such bad pollution that they are obscured for weeks on end from overhead satellite photography.[9] It seems likely that when many Chinese die from pollution and resulting damage to public health, environmental movements will become another form of protest against the regime. Beijing periodically unveils new plans (e.g. Agenda 21 announced in March 1994) but few have any impact as the drive for economic growth destroys any desire for sustainable development.

The environmental problems will have different impacts in different parts of China. If one is blinded by the formal unity of China, one might expect trade between resource-rich and resource-poor provinces to grow. The World Bank has shown that so far, the precise opposite has happened. It is the logic of decentralization and market forces that this should happen, for Australian coal is cheaper for southern coastal provinces than is coal from northern China. As international energy prices change, so will such calculations. And as the structure of the economies of certain provinces rapidly leaps into and beyond newly industrializing country (NIC) status, their needs for energy and their patterns of trade will change, much as it had for existing East Asian NICs. Beijing has made periodic attempts to control energy imports by provinces but in May 1994 it formally abandoned attempts to impose central control on crude oil imports.

If economic growth continues, the international system will also have to adjust, and quickly. In late 1993 China became a net oil importer and is set to become one of the world's largest importers of energy, leading to either major price increases or significant new discoveries of energy. Consider the impact that a rapidly developing Japan and East Asia had on the international oil market and one begins to get a sense of the adjustments that will be needed for a much larger China. Whereas Japan and the NICs seemed disinclined to take an aggressive approach to access to energy resources, China may adopt a more European or American attitude to questions about continuing access to energy, and most notably a greater willingness to use force to defend rights of access (and prices?). Will Chinese energy planners look to Siberia, Southeast Asia, the Gulf? What prices will it pay, and will it trade arms for energy or use arms to take energy?

Managing the fallout from China's greater energy consumption will also pose challenges to neighbours and the wider world. If economic growth remains concentrated along the coast, then those neighbours downwind will suffer (Japan, Korea, Taiwan). If global warming does turn out to be a problem, then China will, rightly or wrongly, be seen as the major marginal cause of damage. China will both have increased leverage over foreigners anxious to control these problems, but will also find itself under greater international scrutiny.

The final aspect of the weakening of authority in China is of course the uncertainty about leadership. An important reason why Beijing has been unable to regain authority is the imminent demise of Deng Xiaoping and the passing of the remnants of the revolutionary generation. This is not just a succession from one revolutionary to another as in the late 1970s, it is the transition to a post-revolutionary leader. Authority, especially in China's personalized system, will almost inevitably be weaker, no matter who succeeds Deng. The fact that this certainty is known even before the old man is put in his grave, means the struggle is more acute before Deng dies.

The most immediate outcome of this reality, apart from the obvious confusion in important aspects of Chinese foreign and domestic policy, is the knowledge that power snatched now, might not have to be relinquished after Deng dies. Thus provinces and/or powerful individuals have little incentive to lean too far to the desires of leaders in Beijing, for no one knows who will succeed. Policy is likely to drift even further and the longer the death watch lasts, the more entrenched the new power holders will become. If foreigners are expecting China to emerge with a strong leader after a while as happened after 1976, they are likely to be disappointed. Just as Western leaders flounder in the post-Soviet morass, so they will find it hard to spot new centres of real authority in a post-Deng China. As authority fragments, so the outside will tend to deal with China in a more fragmented way. This is a challenge that should be welcomed, for it offers many more ways to interest parts of China in closer relations with the outside world. Weaving such webs of interdependence will offer more scope for tying parts of China into the cloth of international society.

EAST ASIA CHANGES SHAPE

Entwining China into webs of interdependence in part depends on the shape of China and in part on the shape of the outside world. East Asia and the wider world may not be changing quite as fast as China, but the transformations are still significant, and complicate any attempt to understand how China might be tied down. The rapidly changing East Asia poses challenges to China, but also provides opportunities.

Perhaps the most obvious trend in East Asia is the rapid increase in trade and financial flows. China's growth rates are impressive, but they are not much different from that shown by Japan and the NICs in earlier decades. The pattern of international trade and finance mutated in reaction to these earlier star performers and it will do the same in response to China. Early phases of growth were in low-wage manufacturing, forcing Europeans and Americans to move out of these sectors or else find new ways of using high technology instead. As the East Asians moved into higher levels of manufacturing, much the same process was seen in the developed world. Service sectors in the developed world grew sharply so that now 'software' is a more important portion of the value of such manufactured products as cars than 'hardware'. Developed Organization for Economic Co-operation and Development (OECD) countries exported services to East Asia and took in far more in manufactured goods.

By the 1990s, East Asian economies began trading relatively more among themselves, and relatively less across the Pacific. Japanese growth rates reached European levels as social and political change followed decades of economic transformation. Are the NICs contemplating a similar process? If so, then the challenge of Chinese growth will reverberate through the region and the global economy in special ways. China now ranks eleventh among merchandise exporters, with likely prospects of rising to number four (now held by France) by the end of the century. China is already at the stage where it runs massive trade surpluses in low-end manufactured goods with the developed world in Europe and across the Pacific. Even Japan shows signs of running a chronic trade deficit with China. In the short run there will be prospects for Chinese manufacturers to suck in higher level manufacturing imports, as well as food and energy. They will certainly need to import services from the developed world. The sources of these imports may be the OECD states, but if the East Asian NICs increasingly move into this sector of the international economy, China may do more of its trading closer to home. But if the NICs get squeezed between a rising China and the OECD states, then resentment of China may mount.

Of course, as Chinese provinces emerge as more independent economic actors, there will be a wider range of possible interactions with 'the Chinese economy'. Richer Chinese provinces are already finding that their wages are too high to compete with hinterland provinces or the likes of Vietnam. They will move into the NIC category quickly, with all the structural changes we have come to see in Association of South East Asian Nations (ASEAN) states or South

Korea. They may try to maintain market share by buying in cheap labour from the hinterland, and with the massive scale of migration now underway, this process can continue for some time. It is not obvious how some areas, most notably the southern coastal provinces can compete with ASEAN states if they move too quickly out of light manufacturing. Clearly there is much scope for a stalling of economic growth as was seen in Latin America thirty years ago. At this stage it is impossible to be sure how the regional economic linkages will develop, but the one thing that is certain is that important parts of China will become increasingly integrated with, and challenged by, the international economy.

Under these turbulent conditions, it is impossible to devise a simple strategy for coping with China in the international economy. Chinese membership in the World Trade Organization (WTO) is likely, but with strings attached. Whether these conditions include human rights is much less important than the forms of economic conditionality. While the Chinese economy will benefit from WTO entry in the sense of lower tariffs, there are newer and perhaps more significant costs to membership in an open trading system. Chinese industry will be asked to abide by specific rules on transparency and non-state interference with markets. Targets will be set with sanctions if targets are missed. Beijing may agree to such terms but find that, as often happens, they are unable to impose conditionality or transparency on provinces and local entrepreneurs. The WTO will work with Chinese regions, much as they work with Australian or Canadian regional governments, to ensure compliance with international rules. If China is to remain in the WTO and avoid constant rows, it will have to tolerate such nuanced policies from the outside world.

Similar challenges are posed by disputes over China's ability to abide by international accords on copyright or textiles. If progress is made in adding accords on labour relations (social chapter) regulations or environmental policy, then deals should be done at the local level as well as with Beijing. Not all regions will be as important, for there is likely to be a greater differentiation within China between those parts that are players in the international market economy, and those that are backwaters not washed by the powerful currents of global money.

The integration of at least parts of China – and it will become the most important parts of China – with the international economy, will create fundamental shifts in the balance of power. China as a whole is likely to run long-term trade surpluses with the developed world. It will run long-term trade deficits with food and energy exporters. Yet these patterns will be confused by the variable roles played by different parts of China. The result will be a China set for long-term trade rows and with a weak hand to play. China will be tied into webs of interdependence and burdened with many international commitments such as to the WTO for better behaviour. China's ability to implement better behaviour will weaken while the need for it grows. If China finds the current levels of penetration by the international community too intense to tolerate, it has not seen anything yet. If it could see the future clearly, it might not want WTO membership.

A second trend in East Asia and the outside world is towards greater democratization. It is often said, especially by authoritarian rulers, that East Asia does not need Western democracy because the norms are alien. Yet it is curious that just as these views were being articulated with increasing panache, the populations of the more developed parts of East Asia evinced a different view. Japan finally changed its ruling party and South Korea elected the former leader of the opposition. Opposition to the Kuomintang (KMT) in Taiwan gained strength and the people of Hong Kong articulated great concern for democratic values, even in the teeth of Chinese opposition. Some even see cracks in the People's Action Party (PAP) grip on Singapore. What is happening?

To say that there is a trend to greater democratization does not necessarily mean that the objective is a Westminster- or Washington-style system. But greater pluralism is certainly on its way, especially for countries that can demonstrate that economic growth does not have to suffer when people win greater freedom. Apparently it is virtually inevitable that when people feel better off and see neighbours with greater freedoms, there is a natural search for more freedom. In Eastern Europe, when there is less prosperity and there is more ethnic division, the desire for strong authority seems to prevail. But most East Asian states are growing more prosperous and few have major ethnic divides.

Chinese leaders have taken some comfort from what they saw as the tendency to authoritarianism in East Asia. They thought prosperity could be obtained without letting go politically and they were supported in this belief by fellow East Asian leaders. China found willing supporters in the UN human rights conference in 1993 when it argued the cause of cultural distinctiveness as a reason to avoid concessions on human rights. China thought it had a wedge to drive through the West.[10] Perhaps it still does, but probably not for long.

The desire to divide political from economic reform looked sensible as the Soviet Union collapsed and its economy imploded. But as the decentralization of the Chinese economy gathered speed, it became clear that the loss of economic authority also meant the loss of some political authority. The linkage should have come as no surprise to good Communists. Compare popular culture in Guangdong and Beijing and one has a clear sense of how political authority begins to change with economic reform and prosperity. It is true that coastal provinces do not yet have élites that openly espouse more liberal political causes, but nuances can be discerned. Transformations in such élites are the last area where change is seen, for these are the people with the largest vested interest in the old system. In the end, these élites are but the thinnest of veneers which no longer hide more liberal political forces.[11]

For the time being, most people in the coastal regions have little interest in politics. Why should they? Their interest is in making money and making money would be messed up by political agitation. Like the hard working people of Japan and the NICs in earlier decades, it often takes several generations before prosperity is sufficiently entrenched for people to afford the luxury of an interest in politics. But the long-term trend does seem clear. And it might not be quite as long term as people think. Just as China's economic growth has progressed at a faster speed in part because it is not blazing a path on its own, so democracy may come faster because there are plenty of role models and a more supportive international system. This discussion is tentative in part because of the fear that everyone must feel about the strength of the massive social forces unleashed in China. The breakdown of social order, the mass migration, the epidemics of crime all suggest that reforms are on a knife edge. Should economic growth cease, the tendency to authoritarian solutions will grow. The alienation in society that is so evident is often channelled into frenzied moneymaking, but it is also channelled into nationalism and a propensity to support authoritarian solutions. Whether it is military rule in Japan or Korea, East Asians have known similar solutions in living memory and at similar stages of development. It may be that China is not yet through the authoritarian stage. Or it may be those parts of China – the richer parts – are through that stage and will reject attempts by some in other parts of the country to impose such rule. These are the risks of continuing decentralization of authority and differential rates and paths of growth. How will the outside world react to a more liberal part of China, and one that is a more important part of the international trading economy, seeking support against authoritarian and rabidly nationalist leaders elsewhere in the country?

More plausibly, at least in the short term, is a China more sensitive and perhaps even more accommodating on human rights. Assuming the state does not collapse, then China will become more decentralized. Not all parts of China will have the same vested interest in confrontation with the outside world about human rights. Richer provinces, if they continue to import cheap labour from poorer provinces, will violate social chapter rights such as did nineteenth-century Europeans and Americans. But they may also be more liberal societies (perhaps not even with Victorian values) less concerned to suppress the rights of political dissidents. The outside world may find it useful carefully to target its criticism both as to issues and parts of China. The richer world might even find new allies in such campaigns if they can cause political changes elsewhere in East Asia. If ASEAN states have been cajoled into more liberal labour practices, they are unlikely to be willing to protect China's ability to undercut their manufacturing industry with cheap and badly treated labour.

These changing attitudes to political rights also mesh with the trend towards greater transnational linkages. The trend, otherwise known as 'globalization', has been sweeping through East Asia much as it has done in other parts of the developed world. This is not the place to rehearse the causes of the process that has brought both the 'broad-casting' of international culture and more specific 'narrow casting' to sub-national groups. Greater flows of people, the media, and culture in general have broken through all countries in East Asia except North Korea. The result is a weakening of the authority of the state and a change in cultural values over time. Some states, such as Singapore, seem more resistant to parts of the process, but then its relatively larger emigration of young professionals reflects the risks of maintaining such a policy for long. These gradual changes in culture and greater knowledge about the outside world can have a far-reaching impact. West German television was a more important cause of the collapse of East European Communism than was CoCom. The challenge facing China is complicated by the fact that the part of global culture that is obviously Western is not necessarily attractive to the Chinese culture. Experience in non-communist East Asia suggests just how complex the assimilation of cultures can be (e.g. Japan). Disney and Coca-colonialism have their impact on China, but arguably Canto-pop music is more important. Overseas Chinese culture, precisely because it is a *mélange* of the West and Chinese, is more potent in pulling parts of China towards the outside world.

Wang Gungwu and others have shown just how complex the overseas Chinese world really is and no justice can be done in this chapter to the subtleties of the problem. But it is true that just as ethnic Chinese communities change when they are in persistent contact with another culture, so that part of China that has thicker contacts with the outside world will also change. The pull of ethnic Chinese is likely to be stronger, but it will still be a pull resented by earlier sources of authority. Coastal China meets the overseas Chinese for reasons of business, but the impact is often felt in political and cultural terms in the long term. Overseas ethnic Chinese who value their human rights and a different way of life, will not be attracted by a poor China, but a poor Chinese will seek to emulate the overseas Chinese.[12] The more such contacts take place, the more difficult it is for mainland Chinese to see the world in simple nationalistic terms. Western culture may be reviled just as many others seek to revel in its pleasures. Revulsion is increasingly heard as a motive for new nationalism as some Chinese return from America resenting its racism and rough and tumble. But it is harder for Chinese to dismiss fellow ethnic Chinese. Thus, as ethnic Chinese gradually become more middle class and different from mainland China, so they will help to pull China in different directions.

While these forces are by far the most important of the transnational linkages, there are also linkages across other Chinese frontiers, most notably in Central Asia. But here, and in China's southwest, the 'pull factors' are far less clear because it is often the Chinese side of the frontier that is the richest.[13] Networks exist that involve increased trade in goods, people, guns and drugs, but there is not much that is attractive on the other side of the frontier. Where there are major problems, for example with Tibetan and Islamic-based nationalism, the causes are much more complex. Beijing worries aloud about these problems, but its grip on reality seems much better than along the coast where it has fewer ways to respond to the challenge. The trend to transnationalism seems set to develop. As more than 50 per cent of trade between OECD countries is carried out within single firms, so the Chinese will be pulled into these networks as they form part of the global economy. As more than 80 per cent of foreign direct investment in China comes from overseas Chinese, the role of kinship will be more destructive to national unity than many imagine. This is not a matter where governments, Chinese, Western, or any other, control the process. These are market and social forces working in what Braudel might have called 'geological time' to change popular perceptions and priorities.

All these forces for change in East Asia help shape the new balance of power in the region. The changes in China fit into a chaos that some like to see as a pattern of multipolarity. A shrinking Russian military presence, a less rapidly shrinking American role and an uncertain Japan are only the most obvious changes since the end of the Cold War. China is the only power with defence spending rising as a percentage of gross domestic product (GDP) and quite dramatically in absolute terms. If

China's economy continues to grow and is measured in purchasing power parity (PPP) terms, then China has the world's third largest defence spending, and rising fast. But does it makes sense to think in such national military terms in East Asia?

The problem is that the answer is yes and no. Yes, because there is little multilateralism in East Asia and virtually none in security terms. East Asians openly admit that they talk the language of military multilateralism in the region to strengthen national security. And no one following the North Korean problem in the past few years can doubt the power of nationalism and xenophobia.

But there are also some international forces that might mitigate such nationalism. Growing economic interdependence is said to be a force for stability as is the habit of dialogue among East Asians, even on security issues. But economic interdependence without more common political goals can be more a cause of conflict than cooperation – witness the 1930s.[14] And it does not take an especially cynical observer to note that security dialogue in the region is often an excuse for inaction on issues of substance. As Europeans know only too well, dialogue in and of itself does not necessarily build confidence – witness the confidence-sapping negotiations in the Second Cold War of the 1980s.

China's behaviour over the South China Sea or Korea does not suggest that it is serious about multilateral security, and without China not much of substance can happen in East Asia. Given the benefits of good relations with South Korea and good grounds to fear the prospects of nuclear proliferation in Northeast Asia, China has remained remarkably aloof from the settlement of the North Korean problem. Some say China is simply on a steep learning curve about multilateralism and arms control, but as time goes by it appears that China is either a terribly slow learner, or else duplicitous. Part of China's suspicion about multilateral security in the region is its concern that Taiwan might somehow slink its way towards international recognition by involvement in such dialogues, and Beijing has a more general suspicion that multilateralism is merely an excuse for Western intervention in China's domestic affairs.[15]

Optimists might draw succour from the fact that arms transfers to East Asia are falling, albeit not as sharply as elsewhere in the world. But such trends mask the more important increase in indigenous production of defence equipment. Richer and more developed East Asians are obtaining more technology transfers than weapon transfers and are developing impressive capabilities. China and Japan have long been major producers of defence equipment, but South Korea, Taiwan and some ASEAN states are making rapid progress. The connections in the technology transfer sphere are often made within networks established as part of involvement in the wider global market economy.[16] And as the Cold War mechanisms for control such as CoCom are replaced by vaguer and less effective systems, these trends will continue. Although China is likely to remain a target of, and a problem for, the remaining international regimes such as the Missile Technology Control Regime (MTCR), it will be less constrained than in the past in obtaining dual-use technology.

Thus it is hard to find trends in the military security of East Asia which encourage much optimism. When such uncertainty about the balance of power appears at the same time as major economic, political and social change in China and the region, 'econophoria' is not a safe bet. Whether it is the obvious risks in North Korea, or the more latent risks of democratization in Hong Kong or Taiwan triggering military tension (e.g. moves to Taiwanese independence), China is the most important (and uncertain) power shaping regional security. Yet there are ways in which the more positive features of political and economic change might be used to help constrain those who would use military power to upset regional and global security. China, the biggest non-status quo power of them all, can be tied down.

WHERE SHOULD WE TIE THE ROPES?

The nightmare in 2020 is a united, authoritarian China with the world's largest GDP, perhaps the world's largest defence spending, and a boulder rather than just a chip on its shoulder. If that is what we face in 2020, then it will be too late to do much about it. The lead time in military technology requires that some serious thinking about 2020 be done in the next five years. The challenge is to make the best of the next generation so that the future is a looser, more pluralist China whose constituent parts are interdependent with different parts of the region. China can be tied in (and down), but even the optimists have to make certain assumptions about the future.

The most important assumption one must make about the future of China is that it will not disintegrate in chaos. This is not the same as saying that China will hang on to a status quo, for it seems increasingly clear that unless power continues to be decentralized in China, the strains on the fabric of the country might become too great. Therefore to say China will avoid chaos, is also to say that it will be a looser sort of political system.

Assumption number two is that East Asia will fail to develop any serious multilateralism. Of course, there will be much talk in the region about the need to work more closely and there will be more multilateral meetings. The ASEAN Regional Forum and the associated CSCAP process will continue. But these meetings are usually an excuse for not undertaking real action that involves some surrender of sovereignty. East Asians will continue to pursue essentially national agendas. In so doing, they will also find it easier not to take a lead on any major foreign policy issue. East Asians will continue to be camp followers, especially on security. If anything is going to be done about the risks of nuclear proliferation or the growth of Chinese military power, the initiative is unlikely to come from within East Asia.

Thus any effort to forge a strategy for dealing with the more complex China in the coming years is unlikely to be led by East Asians, even though they have the greatest stake in the outcome. This odd state of affairs means that it will be particularly difficult to formulate a coherent strategy towards China. The difficulty is compounded by the fact that East Asia and the wider world have conflicting objectives regarding China.

Nevertheless, many of these differences in approach and complexity can be addressed. Given a decent degree of concern about the future shape of China, a strategy of 'positive conditionality' might be composed of at least some of the following features.

- Because China is likely to run a long-term trade surplus with the developed world, those who are major markets (the USA, EU, Japan) will have major leverage. The question is not whether this leverage should be used, but whether it should be used for issues outside of the realm of economic issues. Even if one assumes that trade levers should be used only on trade issues, we will have a major and long-term adversarial relationship with China. While we want China in the WTO, we also can use the terms of entry to ensure that China has an open economy that plays by international rules. Entry to the WTO will provide the quintessential case of positive conditionality, for China can be told that its continuing access to the benefits of the WTO depend on meeting WTO rules, however they are phased in. If China worried about most favoured nation (MFN) issues coming up annually, they will find that WTO entry makes the process permanent, albeit not linked to human rights issues. WTO conditionality can and should be imposed at various levels of the Chinese economy in order to ensure greater compliance. Positive conditionality can be targeted on specific sectors of the economy where violations of trade rules are most serious.

- Because China is likely to be a long-term importer of high technology from the developed world, the West will have major leverage over what China does with its own export of technology. The greatest concerns in the West are Chinese policies on arms sales and nuclear exports and, once again, a strategy of positive conditionality can be applied. Whether it is through the post-CoCom system or related regimes for controlling arms exports, a strategy can be evolved whereby China is offered access to technology in exchange for good behaviour. Chinese arms transfers to Pakistan have already provoked Western sanctions and the experience indicates that a firm Western policy can have an effect on Chinese behaviour.
- Many of the current disputes regarding conditionality and China concern Beijing's human rights policy. A more effective policy would offer the provision of aid if progress is made on human rights. Aid can be targeted to more specific projects that assist the private sector of minority groups. European countries have developed a great deal of experience in such targeted, 'know-how' aid to Eastern Europe and similar projects can be applied to China. China claims to be a poor developing country and therefore worthy of aid, but as parts of China approach NIC levels, aid can be more specifically targeted. From a Western point of view there is certainly little point in providing aid to a country whose human rights policies are abhorrent and not improving.
- Perhaps the easiest areas of positive conditionality concern specifically targeted measures to deal with unsavoury aspects of Chinese growth. Obvious problem areas include Chinese environmental damage, piracy, migration, or the drug trade, and all can be addressed by offering specific assistance. Such aid needs to be targeted precisely and as much as possible as part of multilateral efforts. The idea is to tie China into cooperative practices with its neighbours and the outside world. If the Chinese see the benefit of good behaviour, they are more likely to cooperate, and that good behaviour need not necessarily only be sought at the national level.

China remains unique among the great powers as having little practice in genuinely working multilaterally with the international community. Witness China's behaviour in the United Nations. It was only in April 1994, some 23 years after China assumed its UN seat, that China drafted an important Security Council statement. By and large China has a remarkably nineteenth-century attitude to sovereignty and it is particularly unsuited to the post-modernism of the twenty-first century. The more that China can see the benefits from genuinely multilateral diplomacy, at least in terms of greater respect, the less paranoid China will feel about the world and the less the world will feel paranoid about China.

Such elements of a strategy of positive conditionality require greater cooperation at the G-7 level. The EU has already demonstrated that Chinese attempts to punish Britain for its Hong Kong policy can be beaten back by a robust warning on the General Agreement on Tariffs and Trade (GATT). Such multilateral cooperation in dealing with China can be extended to other G-7 partners. This is not to say that China should be treated like the Soviet Union during the Cold War. The conditions are very different, if only because of China's greater importance for international trade and the absence of clear divisions into East and West. But precisely because of this more complex picture and China's greater interdependence with the global market economy, there are many more means of tying China in (and down). The purpose of such a strategy is to create a China that is more accustomed to multilateralism, to engaging in substantial dialogues, and to real interdependence in social as well as economic terms. The more that China is tied into these relationships, the less heavy the chip on its shoulder will become, and the less ready it will be to use force to settle disputes. Relations with China will naturally be adversarial in many important respects, but the hope must be that by constraining its behaviour, the worst excesses of adversarial behaviour – military tension – can be avoided.

NOTES

1 William Overholt, *China the Next Economic Superpower*, London: Weidenfeld, 1993; Harry Harding, 'The Prospects for China', *Survival*, vol. 36, no. 2, Summer 1994. See also James Cable and Peter Ferdinand, 'China: Enter the Giant' *International Affairs*, vol. 70, no. 2, Spring 1994.
2 For the pessimistic fashion see Richard Hornik, 'Bursting China's Bubble' and Gerald Segal, 'China's Changing Shape', both in *Foreign Affairs*, vol. 73, no. 3, May 1994. See also Gerald Segal, *China Changes Shape*, London: Brassey's for the IISS, Adelphi Paper no. 287, March 1994.
3 Details in note 2 as well as *China News Analysis*, no. 1508, 15 April 1994.
4 Anjali Kumar, 'Economic Reform and China's Internal Market', *Pacific Review*, vol. 7, no. 3, 1994.
5 Richard Margolis to a *Financial Times* conference on capital markets, London, 28 April 1994, *The Economist*, 30 April 1994, p. 97. See also Nicholas Lardy, *China in the World Economy*, Washington: Institute of International Economics, 1994.
6 Half the UK population moved to the cities within a generation during the Industrial Revolution.

7 *Fazhi Ribao*, 23 February 1994 in FE/1976/G/10 and *Cheng Ming*, no. 198, 1 April 1994 in FE/1976/G/8.
8 Paul Smith, 'The Strategic Implications of Chinese Emigration', *Survival*, vol. 36, no. 2, Summer 1994.
9 These issures are discussed in two articles in *The Pacific Review*, vol. 7, no. 2, 1994.
10 Chen Jie, 'Human Rights: ASEAN's New Importance to China' and Philip Baker, 'China: Human Rights and the Law', both in *Pacific Review*, vol. 6, no. 3, 1993.
11 David Goodman, 'Guangdong' in David Goodman and Gerald Segal (eds), *China Deconstructs*, London: Routledge, 1994.
12 See various chapters in Lowell Dittmer and Samuel Kim (eds), *China's Quest for National Identity*, Ithaca, NY: Cornell University Press, 1993. See also a special issue of *The China Quarterly*, 1994, on the components of Greater China.
13 Roland Dannreuther, *Creating New States in Central Asia*, London: Brassey's for the IISS, Adelphi Paper no. 288, 1994.
14 Barry Buzan and Gerald Segal, 'Rethinking East Asian Security', *Survival*, vol. 36, no. 2, 1994.
15 Paul Evans, 'The CSCAP Process', *The Pacific Review*, vol. 7, no. 2, 1994.
16 These issues are discussed in 'Proliferating Arms in East Asia', *Strategic Survey, 1994–1995*, London: Brassey's for the IISS, 1994.

Index

DATE DUE